CHOICES
SCIENCE LEVEL SIX

Science Notebook

purposeful design
publications

Colorado Springs, Colorado

Purposeful Design Publications is the publishing division of the Association of Christian Schools International (ACSI) and is committed to the ministry of Christian school education, to enable Christian educators and schools worldwide to effectively prepare students for life. As the publisher of textbooks, trade books, and other educational resources within ACSI, Purposeful Design Publications strives to produce biblically sound materials that reflect Christian scholarship and stewardship and that address the identified needs of Christian schools around the world.

References to books, computer software, and other ancillary resources in this series are not endorsements by ACSI. These materials were selected to provide teachers with additional resources appropriate to the concepts being taught and to promote student understanding and enjoyment.

Unless otherwise identified, all Scripture quotations are taken from the Holy Bible, New International Version® (NIV®), © 1973, 1978, 1984 by International Bible Society. All rights reserved worldwide.

Cover
Macrophage, lymphoctyes, and human red blood cells image, © Dennis Kunkel Microscopy, Inc.

Peanut plant, **1.6A**, NCDA
Plankton, **9.5A**, U.S. EPA Great Lakes National Program office
Smarties® is a trademark of the Ce De Candy, Inc., which does not sponsor, authorize, or endorse this textbook.
STYROFOAM® is a trademark of the Dow Chemical Company, which does not sponsor, authorize, or endorse this textbook.

Printed in the United States of America
16 15 14 13 12 11 10 09   1 2 3 4 5 6 7

Science, level six
Purposeful Design Science series
ISBN 978-1-58331-220-9 Science Notebook   Catalog #7520

Purposeful Design Publications
A Division of ACSI
PO Box 65130 • Colorado Springs, CO 80962-5130
Customer Service: 800-367-0798 • www.acsi.org

Name _____

Button Down

Your mom asked you to get her sectioned container of buttons because she needed to sew a button on your dad's shirt. As you were bringing it to her, you tripped over a toy left on the floor. The container flew out of your hands and flung open. Buttons were strewn everywhere. Your job is to pick up the buttons and sort them out so that they can be put back into the proper spaces.

Procedure:
- Get the materials from your teacher.
- Observe the mixture of buttons.
- Divide the buttons into two or three large groups based on their physical characteristics.
- Divide each large group into smaller groups until there are nearly identical buttons in each group.
- Answer the following questions regarding how you sorted the objects.

1. What characteristics did you use to divide each large group?

    Group 1: _____

    Group 2: _____

    Group 3: _____

2. What subdivisions did each large group have?

    a. _____ **b.** _____ **c.** _____

3. How many final groupings did you have? _____

4. Describe the group that has the most buttons. _____

_____

5. Describe some characteristics of the final groupings. _____

_____

_____

6. Describe a button that was so unique it could not be categorized by any group.

_____

7. Fill in the following button classification diagram:

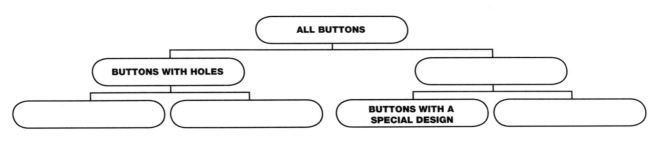

Name _____

Setting Up Shop

Copy the list of supplies from the board onto a piece of notebook paper. You may add any other items. Organize this list into categories as if you were arranging a school supply aisle in a store. Start with a general category and work down to specific groups. Use the concept map below to help you organize your categories. Write the final plan for your aisle in the concept map.

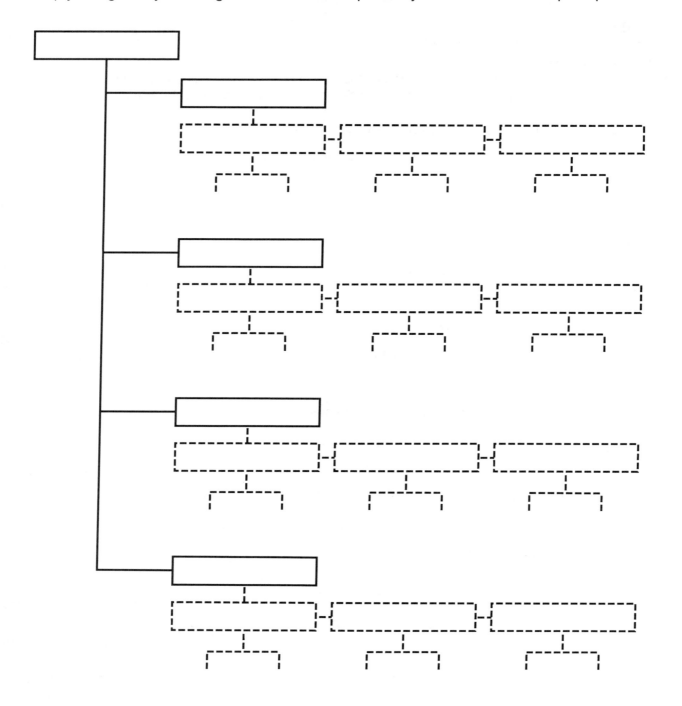

Name _____

Bacteria and Archaea

Fill the diagram with details about bacteria and archaea.

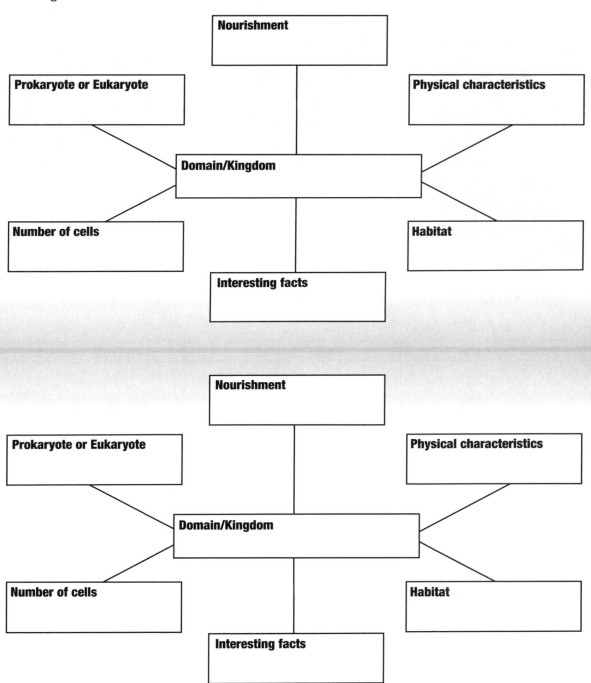

Nourishment

Prokaryote or Eukaryote

Physical characteristics

Domain/Kingdom

Number of cells

Habitat

Interesting facts

Nourishment

Prokaryote or Eukaryote

Physical characteristics

Domain/Kingdom

Number of cells

Habitat

Interesting facts

Write a sentence comparing bacteria and archaea. Include two major features.

_____

_____

Name _____

Microbe Detective

Identify these common genera of bacteria by shapes—bacilli, cocci, or spirilla. Write the scientific name of each bacterium in the appropriate box below. Answer the questions, using your textbook as needed.

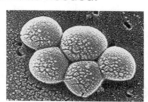

Staphylococcus
resistant to penicillin

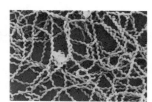

Leptospira
causes fever

Fusobacterium
helps cause gum disease

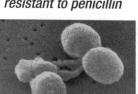

Streptococcus
turns milk into yogurt

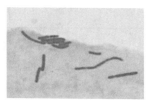

Lactobacillus
produce vitamins in colon

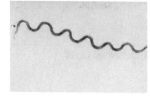

Spirochete
linked to ulcers and cancer

Cocci bacteria	Bacilli bacteria	Spirilla bacteria

1. What is the name of a harmful bacterium? Explain what it does.

_____

_____

2. What is the name of a bacterium that is useful? Explain why.

_____

_____

Name _____

Protists and Fungi

Fill the diagram with details about protists and fungi.

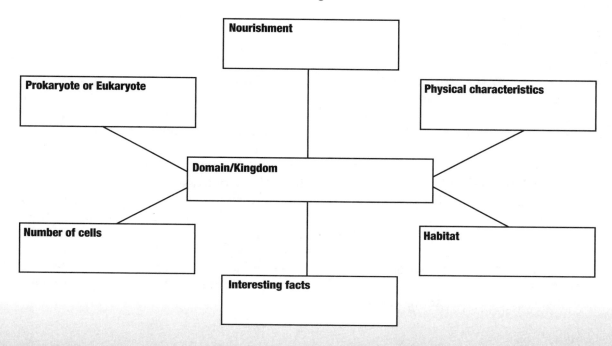

Nourishment

Prokaryote or Eukaryote

Physical characteristics

Domain/Kingdom

Number of cells

Habitat

Interesting facts

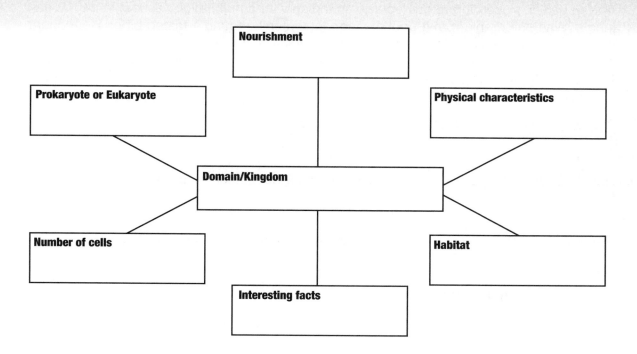

Nourishment

Prokaryote or Eukaryote

Physical characteristics

Domain/Kingdom

Number of cells

Habitat

Interesting facts

Name _____

Cured by Accident

Carefully read the following stories. Answer the questions.

Penicillin

Dr. Alexander Fleming's identification of penicillium mold is one of the best-known stories of accidental medical discoveries. In the summer of 1928, Dr. Fleming was researching bacteria and viruses. He noticed that some mold had contaminated a staphylococcus bacteria culture in one of his petri dishes. Fleming remembered that six years earlier he had accidentally shed a tear into a bacteria sample. Afterwards, he noticed the bacteria would not grow in that spot, concluding that tears have natural antibacterial properties.

That experience caused him to further examine his current sample contaminated with mold. The bacteria had stopped growing where the mold was. Rather than throw the sample away, Fleming took the time to isolate the specific mold. It belonged to the genus *Penicillium*. After many tests, Fleming realized that he had discovered a nontoxic antibiotic substance capable of killing many of the bacteria that cause infections in humans and animals. His work, which has saved countless lives, won him a Nobel Prize in 1945. The medicine known as *penicillin* comes from the Penicillium mold. Fleming later stated that without learning from his previous experience, he would have thrown the contaminated plate away, as many bacteriologists had done.

Quinine

Quinine is found in the bark of the cinchona [sin· ˈkō·nə] tree. Quinine has been used to treat malaria since the early 1600s. A legend suggests that natives in South America were using it even earlier. According to the legend, an Indian with a high fever was lost in an Andean jungle. When he drank from a pool of stagnant water, he found it tasted bitter. He thought it had been contaminated by a tree he called a *quina-quina*. Instead of dying from malaria, his high fever subsided. He shared his discovery with the villagers and others began to use the tree we know as *cinchona*.

1. What did the accidental teardrop happen to reveal?

_____

2. Why did Dr. Fleming decide to not discard the mold-contaminated petri dish?

_____

3. What did Fleming discover as a result of the mold-contaminated petri dish?

_____

4. What is the source of quinine?

_____

5. What did the man accidentally discover when he drank water contaminated by the tree?

_____

Name _____

Plant Kingdom

Fill the diagram with details about the Plant Kingdom.

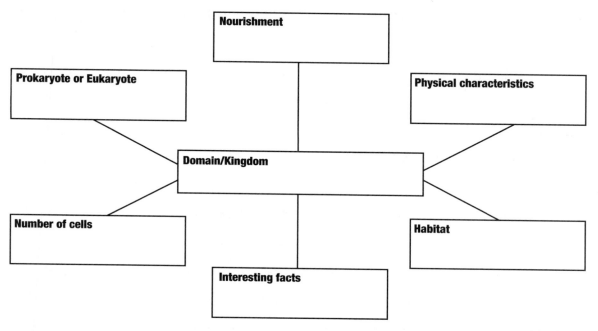

Wayne was asked to make a salad for dinner. The salad contained romaine lettuce, sunflower seeds, broccoli pieces, chopped carrots, asparagus tips, red cabbage, tomato slices, green onion, cucumber slices, dandelion blooms, bean sprouts, and poppy seed dressing. Categorize the ingredients by plant part.

LEAF: _____

ROOT: _____

STEM: _____

FLOWER: _____

FRUIT: _____

SEED: _____

WHOLE PLANT: _____

Name _____

Plant Classification

Listen carefully to the directions the teacher gives about the activity. Then answer the questions.

1. What type of plant transport systems were being illustrated by each team?

Team 1: _____

Team 2: _____

2. Which of the two teams were able to complete the task more quickly and more easily?

_____

3. What made it faster and easier? _____

_____

4. Explain how the transport system affects the height of a plant. _____

_____

_____

Draw examples of a seed, a leaf, and a stem with vascular bundles for a monocot and a dicot.

	Angiosperm monocot	Angiosperm dicot
Seed		
Leaf		
Stem		

Name _____

Dichotomous Key

A dichotomous key provides characteristics to identify unknown plants. Use the dichotomous key below to find the names of the plants pictured. Classify the plants as *angiosperms* or *gymnosperms*. If a plant is an angiosperm, label whether it is a monocot or dicot.

_____   _____   _____

_____   _____   _____

_____   _____   _____

_____   _____

_____   _____

_____   _____

1. a. Leaves are needles or scales. *Go to Step 2*.
   b. Leaves are broad or flat. *Go to Step 3*.

2. a. Seeds are not enclosed, but on scales of a cone. *Pinus strobus L.*
   b. Seeds are small berries. *Juniperus communis*

3. a. Seeds have one cotyledon. *Zea mays L.*
   b. Seeds have two cotyledons. *Go to Step 4*.

4. a. Plant is a tree or a shrub. *Acer rubrum*
   b. Plant is herbaceous or short. *Arachis hypogaea*

Name _____

Pond Scum

Follow the directions and answer the questions.

1. Place 30 mL (1 oz) of pond water in a petri dish.

2. Use the hand lens to observe particles floating in the water.

3. Draw and describe your observations:

   _____

   _____

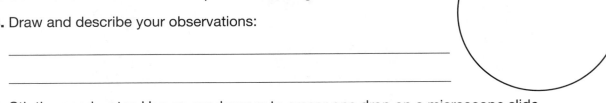

4. Stir the pond water. Use an eyedropper to smear one drop on a microscope slide.

5. Observe this slide under the microscope. Be patient if you do not see anything immediately. Move the slide slightly. Adjust your magnification, as well as the coarse and fine adjustments. If you still do not see anything, repeat Step 4. Describe your observations.

   _____

   _____

6. Your teacher will place five drops of methylene blue stain in the pond water so that the water is blue.

7. Stir the pond water, being very careful because the blue pond water can stain your hands and clothing. Use an eyedropper or pipette to place one drop on a clean microscope slide. Place a cover slip over the drop of water.

8. Observe this slide under the microscope. Describe your observations.

   _____

   _____

9. Illustrate your second observation in the circle to the right.

10. How many cells do the organisms have? _____

11. Compare your two observations. What is different?

    _____

    _____

    _____

    _____

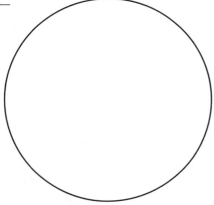

Name _____

Classification Lab

Follow your teacher's instructions to complete the lab and answer the questions.

Question: How is moss different from mold or lichen?

Predict:

_____

_____

Try It Out:

Station 1: Mold

1. Using the hand lens, carefully observe the mold. What color is it? _____

2. Describe other characteristics. _____

_____

3. Is mold a plant? _____ If not, what kingdom does it belong to? _____

4. How does it get its food? _____

Station 2: Lichen

1. Using the hand lens, carefully observe the lichen. What color is it? _____

2. Other than color, describe what it looks like. _____

_____

3. Is lichen a plant? _____ If not, what kingdom(s) does it belong to? _____

Station 3: Moss

1. Using a hand lens, carefully observe the moss. What color is it? _____

2. Using the ruler from the dissection kit, measure how tall the moss is and record. _____

3. Is moss a vascular or nonvascular plant? _____

4. Explain why the moss is not very tall like most other plants. _____

_____

5. Take the dissection needle and carefully pull the moss apart. Observe the top part of the plant and the roots. Describe what you see.

_____

_____

_____

Name _____

Microscopic Study

Carefully look into each microscope. In the circular fields below, draw what you see for each organism. Fill in the blank with what domain and/or kingdom the organism belongs to.

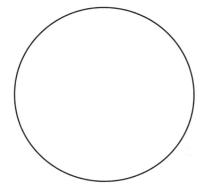

Bacteria
domain/kingdom _____

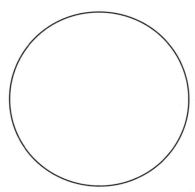

Yeast
domain/kingdom _____

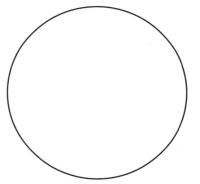

Paramecium
domain/kingdom _____

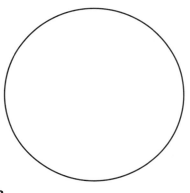

Amoeba
domain/kingdom _____

Analyze and Conclude:

1. What do all of these organisms have in common? _____

2. What type of protist are the amoeba and the paramecium? _____

3. What structure helps the paramecium move? _____ Is it visible? _____

4. What does the amoeba use for movement? _____

5. Describe some visible differences by comparing each of the organisms.

_____

_____

Name _____

Tropism Review

Match the tropism terms with the examples below.

1. thigmotropism   **a.** plants bend or move toward light

2. gravitropism   **b.** plant stems become thicker on one side in order to encircle a pole

3. hydrotropism   **c.** a seed develops and breaks through soil in total darkness

4. phototropism   **d.** tree roots become thicker when closer to a source of water

Pretend you are a plant. Imagine a day in your life. Write an essay describing your climate, your responses to stimuli and sunlight, as well as available water and nutrients.

_____

_____

_____

_____

_____

_____

_____

_____

_____

_____

_____

_____

_____

_____

_____

_____

_____

Name _____

Carrot Investigation

Carrots were cut and placed in water for several days. Write a prediction about how water and sunlight will affect the carrots.

Question: How will water and sunlight affect the carrots?

Predict: _____

Analyze and Conclude:

Keep in mind what you have learned about tropisms. Answer the following questions:

1. What changed in the carrot's appearance?

_____

_____

2. Did all of the carrot tops sprout? _____

3. What does this tell you about how to perform an investigation?

_____

_____

4. Use a ruler to measure the longest sprout and then record the results. Round the measurements to the nearest millimeter (or $\frac{1}{4}$ inch).

_____

5. Was your hypothesis correct? _____

6. List any other interesting observations.

_____

_____

7. When your teacher placed the sprouted carrots away from the window, what happened to the carrots?

_____

8. When the potted plant from the demonstration was placed on its side, what happened?

_____

9. What characteristic of plants did this show? _____

Name _____

Vocabulary Review

Write the number of the word on the blank of its definition.

1. angiosperm

2. autotroph

3. binomial nomenclature

4. dicot

5. eukaryote

6. gymnosperm

7. heterotroph

8. monocot

9. prokaryote

10. taxonomy

11. tropism

12. vascular

_____ a unicellular microorganism without a distinct nucleus

_____ having a system of tubes to transport water and nutrients

_____ a vascular plant whose seeds are not enclosed; most produce cones

_____ an organism that has a nucleus and belongs to the domain Eukarya

_____ an organism that obtains food from other sources

_____ a plant having seeds with two cotyledons

_____ a vascular plant that bears flowers and fruit

_____ the classification system involving two names for each organism

_____ the reaction of a plant to different stimuli

_____ the system of classifying organisms according to characteristics

_____ a plant having seeds with a single cotyledon

_____ an organism that make its own food

Name _____

Chapter 1 Review

Develop the following chart to illustrate the information about domains and kingdoms learned in this chapter.

| _____ |
| (Domain name) |
| _____ |
| (example) |

| _____ |
| (Domain name) |
| _____ |
| (example) |

_____
(Domain name)

_____	_____	_____	_____
(Kingdom name)	(Kingdom name)	(Kingdom name)	(Kingdom name)
_____	_____	_____	_____
(example)	(example)	(example)	(example)

Answer the following questions:

1. Carolus Linnaeus is considered the *Father of* _____.

2. _____ was a lens grinder who observed bacteria using a microscope he invented.

3. Penicillium is a fungus known to kill _____.

4. Give one example for each of the following:

   a. Animal-like protist: _____

   b. Plantlike protist: _____

   c. Fungus-like protist: _____

5. Organisms in the _____ domain can tolerate extreme conditions that most bacteria cannot.

6. Monocots have _____ cotyledon and leaves with parallel veins.

   Dicots have _____ cotyledons and leaves with branching veins.

Circle the correct choice.

7. A conifer is a type of (angiosperm/gymnosperm).

8. A bean plant is a type of (angiosperm/gymnosperm).

Name _____

Match Up

Match the numbers on the continents with the animals that are native to that continent or its coastal areas. You may look these up in an encyclopedia, dictionary, or on the Internet. Some have more than one answer.

_____ polar bear
_____ grasshopper
_____ lion
_____ eagle
_____ slug
_____ oyster
_____ gorilla
_____ rattlesnake
_____ seagull
_____ tree frog
_____ salamander

_____ blue whale
_____ butterfly
_____ sea otter
_____ penguin
_____ worm
_____ conch
_____ ostrich
_____ moose
_____ Komodo dragon
_____ tuna
_____ salmon

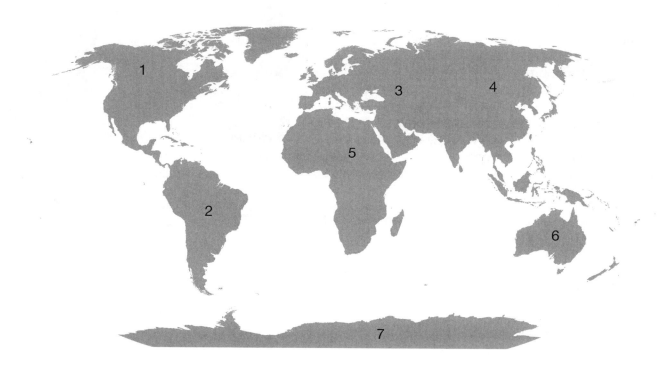

Name _____

The Goose or the Egg

During the twelfth century, monks did not eat meat on holy days. They would, however, eat geese, believing them to be fish. These geese would spend winters in Great Britain, but because explorers had not yet made the discovery, no one knew that the geese nested in the Arctic in the spring.

barnacle geese
(Branta leucopsis)

Barnacles in the coastal waters of Great Britain and Europe look like goose eggs. These eggs have frills, or cirri, which resemble feathers. The cirri are actually wispy hairlike extensions used for capturing food. The barnacles attach themselves to solid surfaces using their peduncles, which are stalklike extensions of their bodies. Some people in the twelfth century thought the peduncles looked like beaks. The Medieval monks even supposed the barnacle eggs attached to logs until the eggs became large enough to fall into the sea and develop into full-grown geese.

Famed sixteenth-century botanist and surgeon John Gerarde claimed he saw birds at different stages of development within the barnacle shells. He chronicled his findings in his book *The Herball or Generall Historie of Plantes*. By the time Carolus Linnaeus assigned names to these animals in the eighteenth century, everyone knew that geese and barnacles were two different animals. Familiar with the myth, Linnaeus named the barnacles to memorialize the tale. *Lepas anserifera* literally means "goose bearing."

goose barnacles
(Lepas anserifera)

1. Which of the two animals mentioned in the article is a vertebrate?

_____

2. Why are the names so similar?

_____

3. Where do barnacle geese live during the spring?

_____

4. Why did the European monks not know this?

_____

5. The monks thought the barnacles' cirri were goose feathers. What are they really?

_____

Name _____

Identifying Beaks and Feet

Birds gather food based on the size and shape of their beaks. Some birds have beaks that are long and tubular. Others have beaks that look like a pouch. Still others have beaks that are shaped more like a drill or a chisel.

Choose members of your group to perform the six different activities—one person per activity. Each person will have 10 seconds to obtain food by using a beaklike tool, while holding the tool with only one hand. Discover how each type of beak functions in obtaining specific types of food. Work together as a group to complete the chart below, using the names of the birds listed, and answer the questions.

| spoonbill | hummingbird | finch | woodpecker | robin | duck |

Station	Food	Amount of Food Obtained	Tool Used	Beak Type	Bird
1					
2					
3					
4					
5					
6					

1. Which utensil did you find most efficient? _____

2. Which utensil did you find most difficult to use? _____

3. Is there a utensil that could be used for more than one food type? _____

Look at the pictures above and describe why each type of bird's foot is suitable for its function.

snatching prey: _____

perching/walking: _____

paddling: _____

Name _____

Rainforest Regroup

Below are the names of some of the animals that live in the rainforests of South America, ranging from tiny insects to large mammals. Put them in the right category.

> **Animals:** piranha, katydid, rhinoceros beetle, anteater, macaw, boa constrictor, spider monkey, sloth, parrot, morpho butterflies, poison arrow frog, black caiman, toucan, jaguar, iguana, leaf-cutter ant, red-eye tree frog, bat, anaconda, eagle, electric eel

Mammals	Reptiles	Invertebrates
_____	_____	_____
_____	_____	_____
_____	_____	_____
_____	_____	_____
_____		

Amphibians	Fish	Birds
_____	_____	_____
_____	_____	_____
		_____
		_____

Give three characteristics of each vertebrate below. Label each animal as an *endotherm* or an *ectotherm*.

fish: _____

_____

reptiles: _____

_____

amphibians: _____

_____

birds: _____

_____

mammals: _____

_____

Name _____

Slug or Worm

Read the article below. Using the information from the article, fill in the Venn diagram comparing a nudibranch and a polyclad flatworm.

Nudibranchs, also known as *sea slugs*, have thousands of known species around the world. These mollusks can be brightly colored or camouflaged so they blend in with their surroundings. Nudibranchs have exposed gills that sometimes can sting. They have a pair of tentacles located on the top of their heads that scientists believe are used to find food or even a mate. All nudibranchs are carnivorous, sometimes even eating other nudibranchs. They move in much the same way as the terrestrial slug moves. Most of them have a muscular foot and hairlike cilia that help them glide across sand. Many also produce very sticky mucus, which enables them to stay stationary in a swift current. Others have the ability to swim in the water.

Polyclad flatworms also live in the ocean. These animals belong to the platyhelminthes phylum and are not related to mollusks. They have tentacles and often have brilliant colors. Their bodies are thin and delicate when compared to nudibranchs. These worms move by using cilia that cover their body. They swim faster than nudibranchs. Most common flatworms are very active carnivorous eaters. They can be a problem to the oyster or clam farming industry. One type of flatworm has an algae growing in it. At low tide, it comes to the surface, where the algae can then use photosynthesis to produce food. The food produced feeds not only the algae, but also the flatworm.

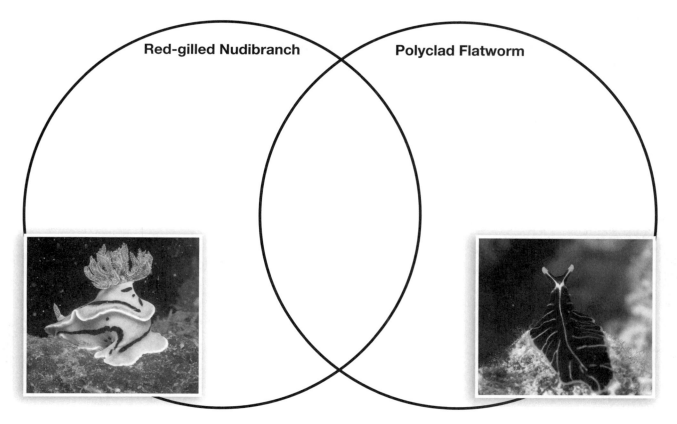

Red-gilled Nudibranch   **Polyclad Flatworm**

Name _____

Classy Clams

Group the following mollusks into their proper class by writing each animal's name under its corresponding shell. Three of the listed animals do not typically have an exoskeleton.

Word Bank	clam	conch	nautilus	octopus	oyster
	scallop	slug	snail	squid	

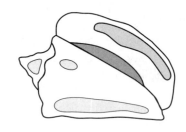

CEPHALOPOD

_____

_____

_____

BIVALVE

_____

_____

_____

GASTROPOD

_____

_____

_____

Identify the arthropods by placing an *A* in front of the arachnids, a *C* in front of the crustaceans, and an *I* in front of the insects.

_____ **1.** scorpion

_____ **2.** brine shrimp

_____ **3.** hermit crab

_____ **4.** tick

_____ **5.** beetle

_____ **6.** barnacle

_____ **7.** black widow

_____ **8.** lobster

_____ **9.** grasshopper

_____ **10.** wood louse

_____ **11.** tarantula

_____ **12.** crayfish

_____ **13.** bumblebee

_____ **14.** blue crab

_____ **15.** cicada

_____ **16.** dragonfly

_____ **17.** mite

_____ **18.** wolf spider

Name _____

Phyla Identifier

Identify the phyla of these marine invertebrates. Label each as *arthropod*, *mollusk*, *echinoderm*, *cnidarian*, or *poriferan*.

squid
_____

sand dollar
_____

pink sea whips
_____

hydroid
_____

arrow crab
_____

hermit crab
_____

barnacle
_____

sea urchin
_____

sea fan
_____

sea cucumber
_____

nautilus
_____

octopus
_____

anemone
_____

nudibranch
_____

spiny lobster
_____

vase sponge
_____

Briefly compare and contrast the three phyla of worms—annelids, platyhelminthes, and nematodes.

_____

_____

_____

Name _____

Mapping Invertebrates

Fill in the concept map below. The central topic is invertebrates. Add the different phyla and the corresponding classes you have learned. Include the following words: _crustaceans, bivalves, echinoderms, gastropods, cnidarians, cephalopods, platyhelminthes, arthropods, nematodes, insects, annelids, poriferans_, and _mollusks_.

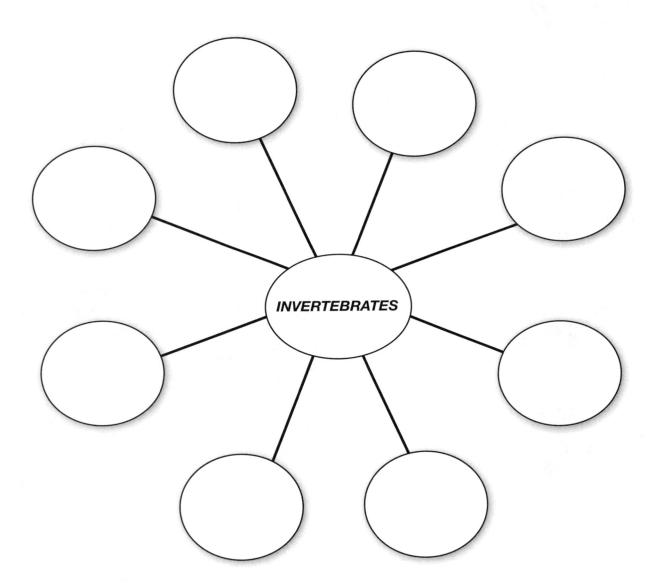

Name _____

Fish Dissection

Your teacher will distribute a fish, a pair of scissors, a dissection knife, a magnifying lens, dissection needles, and gloves to each group. Wait until all the directions have been given. Discuss and answer the questions with members of your team. Let each team member have a turn in the dissection.

Part A: Observation

1. Feel the fish's skin. Describe the color and the texture.

   _____

2. Examine the fins and their location. What function do the bonelike parts of the fins serve?

   _____

3. Remove a scale with the knife. Examine the scale with the magnifying lens and draw it below.

4. Examine the eye. Is there an eyelid? _____ What covers the eye? _____

5. Open the mouth and observe. What color are the gums? _____

   Describe the teeth. _____

6. Carefully examine the tongue. How it is different than a human tongue?

   _____

7. Lift the operculum, which is the flap covering the gills. Use the dissecting knife to gently cut it off as close to the eye as possible. Examine the gills that are now exposed. Describe their appearance, texture, and function.

   _____

   _____

8. Why are fish gills red?

   _____

9. Why do you think the gills have a feather-like appearance?

   _____

10. Label the external parts of the fish on the diagram below.

Name _____

Fish Dissection, continued

Part B: Dissection

1. You are about to remove a rectangular piece of skin and muscle to expose the internal organs. Use the dissection knife or scissors to make a shallow longitudinal incision beginning at the anus and cut from there to a point just beyond the pectoral fins. Be careful not to cut any internal organs.

2. Make two shallow, parallel cuts—one from each end of the incision made in Step 1. Cut almost to the dorsal fins. You should now have three sides of the rectangle finished.

3. Cut the fourth side of the rectangle parallel to your first incision, and gently remove the piece of skin.

4. You should see the liver and intestines. Label them on the diagram below.

5. Using the knife and forceps, dissect the liver from the fish. The liver is part of the digestive system and maintains the correct levels of blood chemicals and sugars.

6. Examine the heart, which lies behind the now-removed liver. The dark red color is from the blood. The remaining organs serve in digestion or reproduction.

7. The heart has two chambers—the atrium and the ventricle. The atrium is on top. Dissect the heart from the fish and make a longitudinal incision through the heart to view the two chambers. Find the stomach and swim bladder. The swim bladder controls the fish's buoyancy. Label the *heart, stomach,* and *swim bladder* on the diagram.

8. Trace the spinal cord to the brain of the fish. Use the knife and forceps to remove tissue as

   needed. Describe the brain of the fish. _____

   _____

9. Follow your teacher's instructions to clean your equipment and your lab station.

10. Finish labeling the diagram and answer the questions.

11. What do you think this fish eats? _____

12. What predators would this fish have? _____

13. My fish is a bony/cartilaginous/jawless fish. (Circle the correct type.)

Name _____

Island Discovery

Dr. Wood and Dr. Shipman have recently explored a new island in the South Pacific. They spent the day looking for signs of animal life. Although they found coastal sea creatures and some seabirds, they found no land animals. Some seashells were interesting to Dr. Wood. When she and Dr. Shipman returned to the laboratory, they identified the shells. To find out their results, look at shell *A* on **BLM 2.6B Seashells**. Follow the steps on the dichotomous key below. Continue until all shells are determined.

1. a. This shell is a univalve. *Go to Step 2.*
 b. This shell is a bivalve. *Go to Step 5.*

2. a. This shell is spiky. *Go to Step 3.*
 b. This shell is smooth. *Go to Step 4.*

3. a. This shell is spiky just at the top. *Strombus gigas* (Queen Conch)
 b. This shell is spiky all over. *Chicoreus dilectus* (Lace Murex)

4. a. This shell is somewhat spherical. *Polinices duplicates* (Shark Eye or Moon Snail)
 b. This shell has a long tapered end. *Ficus communis* (Paper Fig Whelk)

5. a. This shell is ridged. *Go to Step 6.*
 b. This shell is not ridged. *Go to Step 7.*

6. a. This shell is long and rectangular. *Arca zebra* (Turkey Wing)
 b. This shell is fan-shaped. *Chlamys rubidus* (Pink Pacific Scallop)

7. a. This shell is jagged. *Crassostrea gigas* (Pacific Oyster)
 b. This shell is smooth. *Anomia simplex* (Jingles)

Name and give characteristics of the shells.

A. _____

B. _____

C. _____

D. _____

E. _____

F. _____

G. _____

H. _____

Name _____

No Longer Extinct

Read the story and answer the questions below.

In 1836, naturalist Louis Agassiz described a fossil fish he called *Coelacanthus*. This Greek word means "hollow spine." Several things made this fossil look different than modern fishes. The shape of the mouth and teeth indicated the coelacanth usually swallowed its prey whole. The scales were rough and very tightly woven. It almost looked like the coelacanth was wearing armor. The size of the gill indicated the fish lived fairly deep in the water and was less active than some fish. Other fossils were found and examined. All scientists agreed this was an extinct fish. Then, a few days before Christmas 1938, Captain Hendrik Goosen caught a very strange-looking fish in his shark net in the Chalumna River in South Africa. Marjorie Courtney-Latimer, curator of a small local museum, was invited to examine this blue, 1.5 m (5 ft) fish. The fish was eventually identified by a local professor as a *coelacanth*. It was given the scientific name *Latimeria chalumnae* in honor of Ms. Courtney-Latimer.

In July 1998, a brown coelacanth was caught in a deep-water, shark net in Indonesia. This location was 10,000 km (6,200 mi) from the last discovery. The local people were familiar with this fish. They called it a *raja laut*, which means "king of the sea." In September 2003 a coelacanth was caught in the waters off Tanzania. This made Tanzania the sixth country to find a coelacanth. In fact, there have been at least 35 recorded findings in the waters off Tanzania. On July 14, 2007, a fisherman off the coast of Northern Zanzibar caught a coelacanth. Although the coelacanth has incited greater scientific interest and popularity, it was not the first fossil fish to be found alive. The first was the lungfish, which is named for its modified swim bladder. Fossils of the lungfish had been discovered prior to the 1830s when Johann Natterer found a live lungfish in the Amazon.

1. When was the first Coelacanth fossil found? _____

2. How many years elapsed from the time Louis Agassiz first described the Coelacanth fossil and

the first live one was caught? _____

3. What did the local Indonesians call the Coelacanth? _____

4. A _____ is an animal that must share the same

habitat with a Coelacanth.

5. What was the name of the fish that was first discovered to be alive when considered extinct?

_____

Name _____

Biome Classification

Your teacher has handed you a list of biomes and some of the animals that live in them. For each biome, provide two characteristics. Then pick two animals from each biome and give two criteria about each one of them. Criteria helpful in classifying your animal could include number of legs, herbivore/carnivore/omnivore, ectothermic or endothermic, hairy or feathered, etc. Examples are provided below.

	Biome	Biome Description	Animal	Animal Descriptive Criteria
Example	temperate forest	four distinct seasons	earthworm	annelid, no legs
Example	temperate forest	moderate temperature	opossum	nocturnal, omnivore
1				
2				
3				
4				
5				
6				

Name _____

Species Diversity

The research papers about the different members of the cat family have been posted around the room. Please examine each one and then answer the following questions regarding the animals.

1. Which animals, if any, are endangered?

_____

_____

2. Which animals look similar to one another?

_____

_____

3. In what habitats do most cat family members live?

_____

4. Give some examples of the food the cats eat.

_____

_____

5. What is the smallest cat that was researched?

_____

6. What is the largest cat researched?

_____

7. Name three facts you learned about cats while reading the research papers.

a. _____

b. _____

c. _____

Name _____

Vocabulary Review

Pretend you are a biologist working in your office at the Animal Research Center. All the files were mixed up when your assistant pulled some Animal Kingdom files from the drawer in your laboratory. Place the label letter and its description of each animal into its correct phylum file and draw one specimen belonging to that phylum.

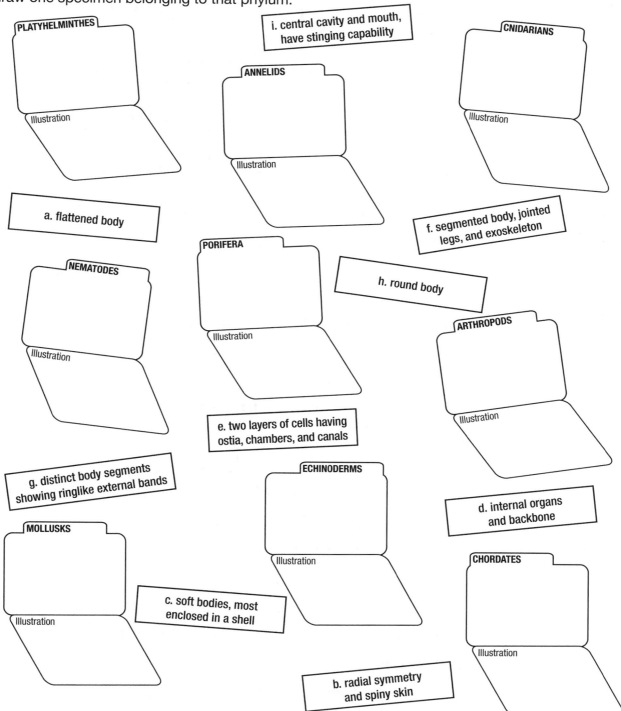

PLATYHELMINTHES

Illustration

i. central cavity and mouth, have stinging capability

ANNELIDS

Illustration

CNIDARIANS

Illustration

a. flattened body

f. segmented body, jointed legs, and exoskeleton

NEMATODES

Illustration

PORIFERA

Illustration

h. round body

ARTHROPODS

Illustration

e. two layers of cells having ostia, chambers, and canals

g. distinct body segments showing ringlike external bands

ECHINODERMS

Illustration

d. internal organs and backbone

MOLLUSKS

Illustration

c. soft bodies, most enclosed in a shell

CHORDATES

Illustration

b. radial symmetry and spiny skin

Name _____

Chapter 2 Review

Answer the questions about animals.

1. List two differences between a frog and a toad.

FROG	TOAD

2. Name three things all mammals have in common.

_____

_____

3. Arachnids are arthropods, but not insects. Name two differences between insects

and arachnids. _____

_____

4. Describe the most significant difference between amphibians and other animals.

_____

_____

5. What is the main purpose of the bony part of a fish's fin and tail? _____

_____

6. Give two examples each of animals with bilateral symmetry and radial symmetry.

_____

_____

7. Three classes of ectothermic vertebrates are _____,

_____, and _____.

8. Name the eight levels of classification from general to specific. _____

_____

Name _____

Egg-speriment

Complete the following exercises:

1. What do you know about eggs? List five details or characteristics.

      a. _____

      b. _____

      c. _____

      d. _____

      e. _____

2. Is an egg unicellular or multicellular? _____

3. Is an egg an organism? _____

4. Does an egg have a cell membrane? _____

5. Explain the function of the cell membrane. _____

_____

6. Describe the physical properties of the cell membrane. _____

_____

7. Compare the characteristics of a cell membrane to the eggshell. Do you think the shell is the

    membrane? Why or why not? _____

_____

    If not, where do you think the cell membrane is located?_____

_____

8. Predict what will happen to the eggshell if the egg soaks in water. _____

_____

9. Predict what will happen to the eggshell if the egg soaks in vinegar. _____

_____

10. If the eggshell is removed, do you think materials can pass in and out of the egg like they do

    in cells? If so, predict how this happens. _____

_____

_____

Name _____

Egg-speriment Data Collection

Fill in the information for each day.

Day 1 Date: _____   Time: _____

Circumference of egg's width before placing in vinegar: _____ cm

Observations: _____

Day 2 Date: _____   Time: _____

Number of hours soaking in vinegar: _____ Circumference: _____ cm

Observations: _____

Predict what will happen when the egg soaks in water. _____

_____

Day 3 Date: _____   Time: _____

Number of hours soaking in water: _____ Circumference: _____ cm

Observations: _____

Predict what will happen after the egg soaks in saltwater. _____

_____

Day 4 Date: _____   Time: _____

Number of hours soaking in saltwater: _____ Circumference: _____ cm

Observations: _____

Predict what will happen when the egg soaks in colored water. _____

_____

Day 5 Date: _____   Time: _____

Number of hours soaking in colored water: _____ Circumference: _____ cm

Observations: _____

Predict what will happen after soaking the egg in fresh water. _____

_____

Day 6 Date: _____   Time: _____

Number of hours soaking in fresh water: _____ Circumference: _____ cm

Observations: _____

_____

Name _____

Ready, Set, Rise

Read and follow the directions below to perform the experiment. Answer the questions.

1. Obtain 5 g (1 tsp) dry yeast, 5 g (1 tsp) sugar, a clear jar or beaker, and 120 ml ($\frac{1}{2}$ cup) warm water from your teacher. Make sure to keep the yeast separate from the sugar.

2. Place the yeast in the clear jar or beaker. Pour the warm water over the yeast.

3. Observe the yeast and water mixture. Record your observations. _____

_____

_____

4. Sprinkle the sugar over the yeast and water mixture. Gently swirl all three ingredients by moving the jar or beaker around in slow circles. Allow the mixture to sit undisturbed for 5–10 minutes.

5. Observe the mixture using your senses of smell and sight. Record your observations below.

_____

_____

_____

_____

6. What function does sugar serve in the above exercise? _____

_____

7. Many baked goods are produced using dry yeast to which water must be added before it becomes active. Why do you think it is important to moisten the yeast using warm water

instead of hot water? _____

_____

8. Explain two ways in which cellular respiration and fermentation differ. _____

_____

_____

9. How are the two processes alike? _____

_____

_____

10. Compare photosynthesis to cellular respiration. _____

_____

_____

Name _____

Fermented Foods

Read the paragraph. Answer the questions that follow.

Generally speaking, fermentation changes a carbohydrate, such as glucose, into an acid or alcohol. Yeast is often used to convert sugar into alcohol. Certain bacteria can change sugar into lactic acid. Including fermented foods in the diet has some health benefits. They can help reduce high cholesterol levels in the blood and strengthen and support the digestive and immune systems.

Foods, such as cucumbers, eggs, mushrooms, and jalapeños can be pickled through lactic acid fermentation. The food item is submerged in a vinegar and saltwater solution. Over time, bacteria grow and digest the sugar stored in the food, producing the tart-tasting lactic acid. Yogurt is formed by adding special bacteria called *L. acidophilus* and *L. bulgarius* to milk. Kefir, a fermented milk drink, is made by adding both yeast and bacteria. Cheese is also made by the conversion of sugar to lactic acid. Bacteria in the milk digest lactose, a milk sugar, producing lactic acid. The acid acts with an added enzyme to curdle the milk, forming curds and whey. Loosely packed curds are often eaten as cottage cheese. After the whey is drained and the curds are compacted, various microorganisms ripen it into cheese. The longer the cheese ages, the more flavor will develop. Other flavors can also be introduced by adding smoke, soaking the cheese in liquids, or adding herbs and spices. Soy sauce is a liquid often used as an Asian food seasoning. This sauce is produced from the fermentation of soybeans, wheat grain, water, salt, and yeast. Soy sauce is known for its salty flavor.

1. What happens during the process of fermentation? _____

_____

2. Which microorganism is used in alcohol fermentation? _____

In lactic acid fermentation? _____

3. How are foods pickled? _____

_____

4. Which type of fermentation is used to make cheese? _____

5. How are block cheese and cottage cheese related? _____

_____

6. List the ingredients used to make soy sauce. _____

_____

7. Predict which type of fermentation is used to make soy sauce. Explain your prediction.

_____

_____

Name _____

Defusing Diffusion

Answer the following questions:

1. What does *selectively permeable* mean? _____

_____

2. How is this characteristic beneficial to the cell membrane? _____

_____

3. Compare and contrast passive and active transport. _____

_____

_____

_____

4. List three types of passive transport.

a. _____ b. _____ c. _____

5. Describe two situations in which active transport would be required.

a. _____

b. _____

6. Explain how diffusion occurs. _____

_____

_____

7. Draw a *Before* and *After* picture of diffusion where the higher concentration of molecules starts on the inside of the cell.

Before	After

8. What is the name of the process in which water moves into and out of cells? _____

How is this similar to diffusion? _____

_____

Name _____

Practicing Transport

Carefully observe your teacher perform some demonstrations. Follow directions and answer the questions.

1. After your teacher sprays the scented air freshener, raise your right hand as soon as you smell it. Observe when your classmates raise their hands. Does everyone smell the air freshener at the same time? Why do you think this is? _____

   _____

2. What process is taking place? _____

3. Explain specifically what is happening. _____

   _____

   _____

   _____

4. Raise your left hand when you smell the vinegar. Which did you smell first, the air freshener or

   the vinegar? Explain why. _____

   _____

5. What process enabled you to smell the vinegar? _____

6. Give another example of diffusion in your home. _____

7. Observe your teacher add food coloring to some water. What happens? _____

   _____

8. Is this an example of osmosis? Why or why not? _____

9. Was any energy required for the movement of molecules in those demonstrations? _____

10. Watch while your teacher performs one more demonstration. Was energy required to move

    the bead "molecule" from one end of the straw to the other? _____

11. What type of transport does this represent? _____

12. What role does the straw play in this demonstration? _____

13. Give a specific example of when this particular type of transport would occur.

    _____

    _____

    _____

Name _____

Cellular Division

Read and complete the exercises below.

1. Liver and brain cells are _____ cells.

2. An egg is a type of _____ cell called a _____.

3. Explain the main difference between body and reproductive cells. _____

_____

_____

4. In the space below, draw a duplicated chromosome. Label the chromatids and
the centromere.

5. List the three stages of the cell cycle in the proper order.

    a. _____   **b.** _____   **c.** _____

6. List the four phases of mitosis in order, and briefly describe the events that take place in
each phase.

    a. _____

_____

    b. _____

_____

    c. _____

_____

    d. _____

_____

7. Gametes are formed during the process of _____.

8. Give three ways in which meiosis is different from mitosis.

    a. _____

    b. _____

    c. _____

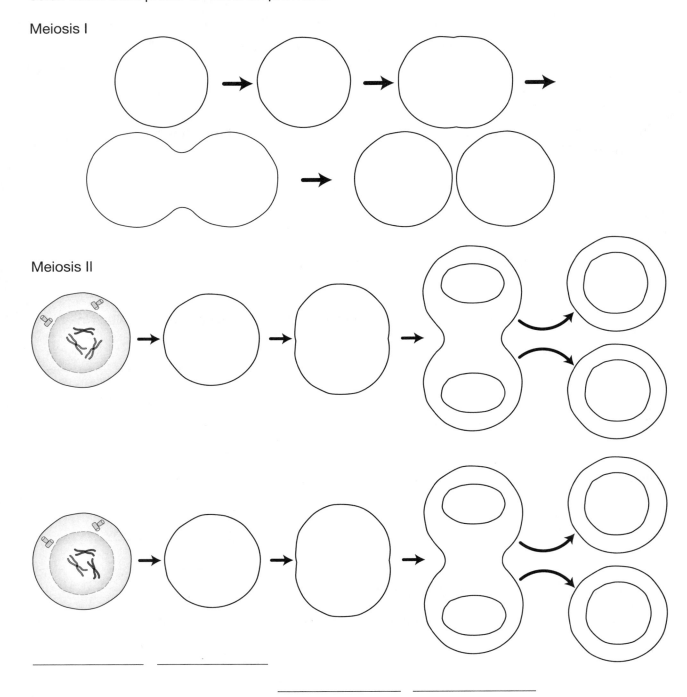

Name _____

Meiosis

Diagram meiosis I and II in the blank circles below. Show at least three pairs of chromosomes in meiosis I. Color the paternal chromosomes one color and the maternal chromosomes a different color. Label each phase on the lines provided.

Meiosis I

Meiosis II

Name _____

Egg-speriment Analysis

Use the results and information recorded on **Science Notebook 3.1B Egg-speriment Data Collection** to complete the following exercises:

1. What was the main process taking place during this experiment? Explain.

_____

_____

2. Did any other process occur during one or more of the daily experiments? If so, explain.

_____

_____

_____

3. Explain why the egg stayed the same size on Day 6, but the water in the container

became colored. _____

_____

_____

4. Compare each of your predictions with the actual results. When were your predictions correct? When were they not?

_____

_____

_____

_____

_____

_____

5. How did this experiment help you understand diffusion and osmosis?

_____

_____

_____

6. List two things you learned from the experiment.

a. _____

b. _____

Name _____

Mitosis in Action

Read **BLM 3.6A Directions: Mitosis in Action**. When your teacher gives you your chromosome number and letter, find your partner. Follow the directions from your teacher to complete the activity. Check the boxes that apply for each phase of mitosis in the chart below. Also, explain what occurs during each stage in your own words. Be specific. You may need to include details that were not acted out in the activity in your explanations.

Stage of Cell Division (Mitosis)	DNA replicates.	Meet at center of room.	Gather to opposite poles of the room.	Use your own words to describe cell/chromosome activity during this stage of mitosis.
prophase				
metaphase				
anaphase				
telophase				

1. What structure did the yarn represent? _____

2. What structure did each student represent? _____

3. Each student pair represented what structure? _____

4. What structure that was not included in the activity is needed for chromatids to separate?

_____

5. If you were to do this activity for meiosis, explain what you would have to do for meiosis I.

_____

_____

_____

_____

_____

Name _____

Binary Buddies

Describe the following types of asexual reproduction. Then on the line below each illustration, identify the type of asexual reproduction being represented.

1. budding: _____

_____

2. vegetative reproduction: _____

_____

3. binary fission: _____

_____

4. spores: _____

_____

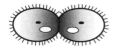      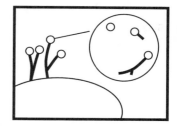

_____   _____   _____

_____   _____   _____

Name _____

Organism Reproduction

Research the organisms listed in the chart below. Fill in the second column with the most common type of asexual reproduction that the organism undergoes to produce offspring. Fill in the third column with details about how the particular organism reproduces in that manner.

Organism	Type of Asexual Reproduction	Description
bacterium		
sea squirt		
spider plant		
paramecium		
tulip		
mold		

Name _____

Vocabulary Review

Fill in the blanks with the term that best completes the sentence. Refer to the Word Bank. Use each term only once.

Word Bank	mitochondrion	budding	glucose	vacuole	binary fission
	fermentation	ribosome	somatic cell	meiosis	diffusion
	gametes	chromatid	endoplasmic reticulum		osmosis

1. A plant cell usually has one large water _____.

2. The _____ of molecules is the reason that odors spread around a room.

3. A _____ undergoes mitosis.

4. Protein production occurs in the _____.

5. The _____ is the organelle in which cellular respiration takes place.

6. Water passes through cell membranes by way of _____.

7. When a chromosome duplicates, each half is called a _____.

8. When cells produce energy they use _____ as the main source of food.

9. Unicellular organisms often reproduce by _____, a type of asexual reproduction in which the parent cell splits into two new cells.

10. When oxygen is not present, _____ takes place in the cells to produce the energy they need for life processes.

11. Materials are transported around the cell by the _____.

12. When a small part of an organism grows outward and eventually becomes an independent organism, it is called _____.

13. _____ are required for sexual reproduction to take place.

14. Reproductive cells are formed by the process of _____.

Name _____

Chapter 3 Review

Complete the following exercises:

1. How many cells are formed following mitosis? _____ Meiosis? _____

2. List the phases of mitosis in order. Next to each phase describe the major events that occur.

    a. _____

    b. _____

    c. _____

    d. _____

3. Use the diagram at the right to answer the following questions:

    a. In which direction will salt move? _____

    b. In which direction will water move? _____

    c. What process moves the salt? _____

    d. What process moves the water? _____

—animal cell

● = salt

water and salt solution

4. Explain what happens during meiosis I. _____

_____

_____

5. Compare and contrast photosynthesis and cellular respiration. _____

_____

6. Compare and contrast cellular respiration and fermentation. _____

_____

_____

7. What are the ways in which unicellular organisms reproduce? _____

_____

8. List three benefits of asexual reproduction.

    a. _____   **b.** _____

    c. _____

9. When would a cell use active transport instead of passive transport?

_____

_____

Name _____

Genetic Probability Puppies

Today you will create a paper puppy using random distribution of genetic information. Assume that all of the puppies in the class have the same two parents. You will be flipping a coin two times to determine each trait your puppy will possess. Each puppy will have its own unique characteristics, or traits.

Your teacher will provide the following: one coin, masking tape, a black pen or marker, construction paper, and scissors.

Procedure:

1. Place a small piece of masking tape on both sides of the coin, but no longer than the diameter.
2. On one side of the coin, write *A* on the masking tape.
3. On the opposite side of the same coin, write *a* on the masking tape.
4. For each physical feature listed below, flip the coin twice to determine what trait your puppy will inherit.
5. Record the letter on the coin after each flip and then write the trait in the space provided.
6. Use the features determined to draw and assemble your puppy out of construction paper.
7. Compare your puppy to that of another student and answer the questions.

Physical Features

1. Coat Color: AA or Aa = Brown, aa = Yellow

Coin Results: _____ Coat Color: _____

2. Ear Shape: AA or Aa = Pointed Ears, aa = Round Ears

Pointed   Round

Coin results: _____ Ear Shape: _____

3. Ear Size: AA or Aa = Large Ears, aa = Small Ears

Coin results: _____ Ear Size: _____

4. Eye Shape: AA or Aa = Shaped Eyes, aa = Round Eyes

Shaped   round

Coin results: _____ Eye Shape: _____

Name _____

Genetic Probability Puppies, *continued*

5. Body Shape: AA or Aa = Potbelly, aa = Hot-dog

Potbelly   Hot-dog

Coin results: _____   Body Shape: _____

6. Whisker Length: AA or Aa = Long, aa = Short

Long   Short

Coin results: _____   Whisker Length: _____

7. Nose Shape: AA or Aa = Pug, aa = Round

Pug   Round

Coin results: _____   Nose Shape: _____

8. Tail Length: AA or Aa = Short, aa = Long

Short   Long

Coin results: _____   Tail Length: _____

9. Sketch the features of your puppy below. Use the sketch to create the puppy using the materials provided by your teacher.

10. Compare your puppy to one other person's. Are all the traits of your puppy the same as that

of your classmate's? _____

11. Which traits are the same? _____

12. Which ones are different? _____

Name _____

Dominant and Recessive Alleles

The term *allele* refers to the variations of a specific gene for a specific trait. Mendel understood that one factor, or allele, for each gene was donated by each of the two parents. These alleles occur in pairs within the offspring. They determine the traits that the offspring displays.

Write all the possible genotype codes next to each trait below. Assign a letter for each trait, using the first letter of the dominant trait name. A capital letter denotes the dominant allele, and a lowercase letter denotes the recessive allele. For example, plant height would be *T* for dominant (tall), and *t* for recessive (short).

Traits Controlled by Dominant Alleles	Dominant Genotype(s)	Traits Controlled by Recessive Alleles	Recessive Genotype(s)
black fur in rabbits		white fur in rabbits	
round peas		wrinkled peas	
green pea pods		yellow pea pods	
yellow-skinned summer squash		green-skinned summer squash	
detached earlobes		attached earlobes	
brown eyes		non-brown eyes	
straight thumbs		bent thumbs	
smooth chin		cleft chin	

Use the terms *dominant, recessive, purebred, hybrid,* and *allele* in a paragraph to describe at least two of the above traits. Use the terms accurately to show your understanding of them.

_____

_____

_____

_____

_____

_____

_____

_____

Name _____

Class Traits

Your teacher will display a transparency. Observe the different traits. Dominant traits are listed in the left-hand column. Survey your classmates for the following traits and write in the number of students displaying each one. After collecting your data, convert your data to percentages. For example, if 15 out of 30 students show brown eyes, record that information as 50%.

Thumbs:

Total students _____

Number with straight thumbs _____

Percentage with straight thumbs _____

Total students _____

Number with bent thumbs _____

Percentage with bent thumbs _____

Eyebrows:

Total students _____

Number with broad eyebrows _____

Percentage with broad eyebrows _____

Total students _____

Number with slender eyebrows _____

Percentage with slender eyebrows _____

Eye Color:

Total students _____

Number with brown eyes _____

Percentage with brown eyes _____

Total students _____

Number with non-brown eyes _____

Percentage with non-brown eyes _____

Summarize the results above using complete sentences. Were there more dominant or recessive traits in your class?

_____

_____

_____

_____

_____

_____

_____

_____

Name _____

Punnett Squares

Follow the instructions to complete four genetic crosses, and then answer the questions. The first cross has been started for you. Brown hair color is dominant over red. The code for brown is B, and the code for red is b.

	b	**b**
B	Bb	
B		

1. Cross a homozygous brown with a homozygous red.

2. What are the genotypes of the offspring? _____

3. What percentage of offspring have brown hair? _____

4. What percentage have red hair? _____

5. Using the same trait as above, cross two heterozygotes.

6. What are the genotypes? _____

7. What percentage of offspring are expected to have brown hair? _____

8. What percentage of offspring are expected to have red hair? _____

9. Using the same trait as above, cross a heterozygote with a homozygous red.

10. What are the genotypes? _____

11. What percentage of offspring are expected to have brown hair? _____

12. What percentage of offspring are expected to have red hair? _____

13. Using the same trait as above, cross a heterozygote with a homozygous brown.

14. What are the genotypes? _____

15. What percentage of offspring are expected to have brown hair? _____

16. What percentage of offspring are expected to have red hair? _____

17. If the offspring of a cross was heterozygous, which trait would appear in the phenotype?

_____

18. Is it possible to have a heterozygous red? Why or why not? _____

_____

_____

Name _____

Pedigree Assessment

Evaluate the pedigree below. A square indicates a male; a circle represents a female. The horizontal line connecting a male and female indicates that these are parents of children. A vertical line from the pair indicates their offspring. If there are multiple offspring from one set of parents, they are indicated below by vertical lines attached to a horizontal line. Individuals with their symbol filled in are affected by a certain gene.

Genes located on the either of the sex chromosomes are called *sex-linked genes*. Color blindness is a sex-linked trait controlled by a recessive gene located on the X chromosome. Since a male's sex-determining chromosome pair is shown as XY, he will be color-blind if he inherits the color-blind allele. This is because there is no corresponding dominant allele on the Y chromosome to counteract the recessive allele on the X chromosome. On the chart below, the filled-in shapes represent an individual that has the color-blind trait. Use the chart and this information to answer the questions.

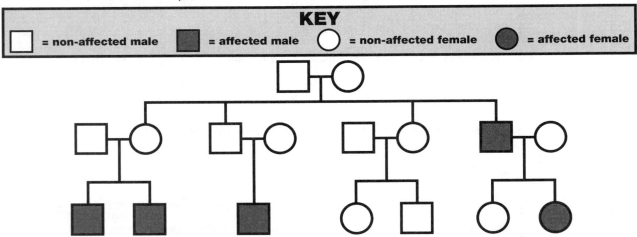

KEY

☐ = non-affected male   ■ = affected male   ◯ = non-affected female   ● = affected female

1. Circle the offspring who express, or display, the color-blind trait.

2. A carrier is someone who does not express, or show, a trait, but can pass it on to his or her offspring. Listen to your teacher's instruction, and then shade one half of the symbol for each of the individuals who are carriers of the trait.

3. Can a male be a carrier for color blindness? Explain. _____

_____

4. How many females are carriers? _____

5. For the family on the right-hand side of the chart, why does one of the daughters produced

    have the expressed trait, but the other does not? _____

_____

_____

6. What would change in the pedigree if the trait of color blindness were dominant?

_____

Name _____

Codominance

Read each example below. Write *codominance* or *incomplete dominance* in the blank lines below to indicate what each example represents.

1. Cows can be black, white, or white with black spots. _____

2. Flowers can be red, white, or pink. _____

3. A chicken can have brown feathers, white feathers, or some of each. _____

People have different blood types based on the presence or absence of certain proteins. The alleles that provide the coding for blood type are I^A, I^B, and i for blood type O. Blood types A and B are codominant, and O is recessive. The O blood type indicates the absence of the A and B alleles. Each parent donates one blood type allele to each offspring.

4. What genotypes are possible for the following blood types?

Type A: _____ or _____   Type AB: _____

Type B: _____ or _____   Type O: _____

5. Complete the cross to show that the father is homozygous for type B blood and the mother is codominant type AB.

6. What are the possible genotypes for the children of these parents?

_____

7. What blood type(s) could their children express?

_____

8. Is there any blood type that could not be expressed by their children?

If so, what is it? _____

_____

9. Complete the cross to show that the mother and father are both heterozygous for type A blood.

10. What are the possible genotypes for the children of these parents?

_____

11. What blood type(s) could their children express?

_____

12. Is there any blood type that could not be expressed by their children? If so, what is it?

_____

Name _____

Incomplete Dominance Puppies

Consider the paper puppies from **Science Notebook 4.1A–B Genetic Probability Puppies** made at the beginning of this chapter. The possible features for each puppy were coat color, ear shape, ear size, eye shape, body shape, whisker length, nose shape, and tail length. For each trait there were two possible outcomes. This is an example of typical dominance in which only one trait fully appears in the offspring. Imagine, however, that the puppy expresses a blend of its parents' traits.

1. Draw the two parent features, and then describe and/or draw what the incomplete dominant blend could look like. Refer to Science Notebook 4.1A–B for the trait options.

	Trait 1	Trait 2	Incomplete Dominant Trait
coat color			
ear shape			
ear size			
eye shape			
body shape			
whisker length			
nose shape			
tail length			

2. Complete the following for a body shape of an incomplete dominant trait. Cross a male dog with a potbelly shape, with a female dog that is hot-dog shaped. Use *P* to represent the potbelly genotype, and *H* to represent the hot-dog genotype.

3. What is the resulting genotype of this cross?

_____

4. Describe the phenotype.

_____

_____

5. In your own words, describe incomplete dominance. _____

_____

Name _____

The Missing Pen

Someone has stolen your teacher's pen! It is up to you and your team to evaluate the evidence to determine who the culprit is. At each station there is a set of data to analyze. Record your findings and then compare it to the list of suspects. The pen thief is the one suspect who matches the evidence in each category.

Hair Sample Station

1. Use the magnifying glass to view the hair samples. Compare them to the one collected at the scene. Write your observations.

_____

_____

_____

2. What are the possible genotypes, or genetic coding, for the color of each sample? Compare it to the one collected at the scene, and place a star next to the one(s) that match. Use the following allele codes for the different hair colors: *B* for dominant brown or black; *b* for recessive blonde or red.

	Sample 1	Sample 2	Sample 3	Sample 4
Color				
Genotype(s)				

Blood Sample Station

3. The thief accidentally cut his/her finger with paper and left a small blood sample. Look through the clues to determine the blood type. Write your observations below. Complete any Punnett squares that may be necessary to come up with a conclusion.

Potential blood types: _____

Suspect matches: _____

Name _____

The Missing Pen, continued

The blood was also evaluated for the color-blindness gene. It was determined that the perpetrator had inherited at least one color-blind allele.

4. If the thief is a female, explain what phenotype she has inherited for color blindness and what

causes each phenotype. _____

_____

5. If the thief is a male, explain what phenotype he has inherited for color blindness and what

causes each phenotype. _____

_____

6. In evaluating the alleles for gender in the genotype, it was determined that the culprit had received an *X* chromosome from the mother and the father. Is the thief male or female?

_____

Pen Sample Station

7. Each suspect was asked to draw a line using the pen that was in his or her pocket. Compare it to the evidence found at the scene. Write your observation of each pen line. Place a star next to the sample(s) that match the missing pen.

Missing Pen: _____

Sample 1: _____ Sample 2: _____

Sample 3: _____ Sample 4: _____

8. Summarize the data by filling in the chart below.

Hair Color	Blood Type	Gender	Color-Blind Allele (Y or N)	Pen Sample Number(s) that Match(es)

9. Review **BLM 4.6B Suspect Profiles** and compare your data to that of the suspects. Who is

the pen thief? _____

10. What other genetic data could have been useful in determining the thief? _____

_____

11. How is genetic data important to law enforcement? _____

_____

_____

Name _____

Genetic Probability Offspring

The puppies you created at the beginning of this chapter are now adults and ready to have puppies of their own. You and a partner will use the genotypes and phenotypes of each of your puppy's traits to determine possible offspring.

Trait	Your Dog's Phenotype	Genotype(s)	Partner's Dog's Phenotype	Genotype(s)
coat color				
ear shape				
ear size				
eye shape				
body shape				
whisker length				
nose shape				
tail length				

1. What are the possible genotypes for the coat color of the offspring? _____

2. What are the possible phenotypes for the coat color of the offspring? _____

3. What are the possible genotypes for the ear shape of the offspring? _____

4. What are the possible phenotypes for the ear shape of the offspring? _____

5. What are the possible genotypes for the ear size of the offspring? _____

6. What are the possible phenotypes for the ear size of the offspring? _____

7. What are the possible genotypes for the eye shape of the offspring? _____

8. What are the possible phenotypes for the eye shape of the offspring? _____

9. What are the possible genotypes for the body shape of the offspring? _____

10. What are the possible phenotypes for the body shape of the offspring? _____

11. What are the possible genotypes for the whisker length of the offspring? _____

12. What are the possible phenotypes for the whisker length of the offspring? _____

13. What are the possible genotypes for the nose shape of the offspring? _____

14. What are the possible phenotypes for the nose shape of the offspring? _____

15. What are the possible genotypes for the tail length of the offspring? _____

16. What are the possible phenotypes for the tail length of the offspring? _____

Name _____

Genetic Probability Offspring, continued

17. Were most of the offspring expressing dominant or recessive traits?

_____

_____

18. Explain why it is possible for some traits expressed in an organism to be dominant, while

others are recessive. _____

_____

19. Compare your results to that of another pair of students. What similarities do you notice?

What are the differences? _____

_____

_____

_____

20. Describe why there are some differences between your results and the results from another

pair of students. _____

_____

_____

_____

21. Describe why there are some similarities.

_____

_____

_____

22. Use the data from the previous page to create a new puppy. Draw it using the phenotypes
that were determined from the cross between your puppy and your classmates'. If there is
more than one phenotype possibility for a trait, choose which one your puppy will inherit.

Name _____

Spot's Heredity

Your teacher will read you a story about a horse named *Spot* and his family. As you hear information about each horse, fill in the horses' names under the column headings and then complete the chart below.

Trait	Spot	Mother	Father	Father's father	Father's mother	Mother's father	Mother's mother
body color							
face color							
leg color							
tail color							
mane color							
eye color							
personality							

Name _____

Spot's Heredity, continued

Answer the following questions about Spot's traits and heredity:

1. Which traits did only Spot have? _____

2. Using the principles of genetics and heredity described in this chapter, how is it possible that Spot could inherit traits that his parents or grandparents did not have?

_____

_____

_____

3. From which relative did Spot inherit the black on his tail and mane? _____

4. Who did Spot inherit his white face from? _____

5. From which family member(s) did Spot inherit his personality? _____

6. Draw a Punnett square for a cross between Spot's parents that accounts for Spot's blue eyes. Use *B* for the dominant brown color and *b* for the blue recessive color.

7. Is Spot's eye color dominant or recessive? _____

8. Draw a pedigree that accounts for Spot's body color. Place Spot at the bottom, his parents one level up, and his grandparents on the top. Make sure to color in all those individuals who have the same body color as Spot.

9. Is Spots' body color dominant, recessive, codominant, or incomplete dominance?

_____

Name _____

Vocabulary Review

Match the following words to the correct descriptions and write them in the blanks below.

Word Bank					
allele	genotype	heterozygous	pedigree	codominance	dominant allele
homozygous	hybrid	phenotype	heredity	recessive allele	incomplete dominance
purebred	trait	Punnett square			

_____ **1.** a characteristic that is inherited

_____ **2.** the passing of traits from parent to offspring

_____ **3.** a different form of a single gene that controls an inherited characteristic

_____ **4.** the allele that hides the effects of another allele

_____ **5.** the allele whose effect is hidden when paired with a dominant allele

_____ **6.** an organism with identical alleles for a particular trait

_____ **7.** an organism that has two different alleles for a particular trait

_____ **8.** a diagram used to show all the possible allele combinations of a genetic cross

_____ **9.** having the presence of two different alleles for the same gene

_____ **10.** having the presence of the same two alleles for the same gene

_____ **11.** a record that shows a pattern of genetic inheritance in a family

_____ **12.** the condition in which both alleles of a gene are actively expressed and neither is dominant or recessive

_____ **13.** the condition in which both alleles are partially expressed in a blended appearance of the trait

_____ **14.** the observable characteristics of an organism's genotype

_____ **15.** the set of genes that make up an organism

Name _____

Chapter 4 Review

Read each example below. Write the letter *C* next to those which represent codominance, an *I* next to those that have incomplete dominance, and a *D* next to those which have complete dominance.

1. A person has blood type *AB*. _____
2. Flowers can be red, white, or pink. _____

3. Flowers can be red or white. _____
4. A human can have brown or blonde hair. _____

5. A chicken can be all brown, all white, or have brown feathers and white feathers. _____

6. Draw a Punnett square for a cross between a mother who is heterozygous for brown hair and a father who is homozygous for blonde hair. Evaluate the Punnett square to determine the probability that a child will have brown hair or blonde hair. Let *B* represent the dominant brown-hair allele and *b* the recessive blonde-hair allele.

_____ % brown hair

_____ % blonde hair

7. How is it possible for two parents with a phenotype showing a dominant trait to produce a child with a phenotype showing the recessive trait? _____

_____

8. *Genetics* is defined as the "study of heredity." Explain what this means. _____

_____

9. Describe Gregor Mendel's pea plant experiments and list two conclusions he came to.

_____

_____

_____

10. What is the difference between a purebred and a hybrid? _____

_____

11. How can the use of inbreeding be beneficial to animal breeders? _____

_____

12. How can the use of inbreeding cause problems? _____

_____

Name _____

The Periodic Calendar

1. What would be the next four characters in the following sequences?

51769342118517693421185176934211851 7 __ __ __ __

JIJCPQRMNDKMNJIJCPQRMNDKMNJIJCPQRMN __ __ __ __

2. The way you found your answers was to find patterns in the numbers and letters. Things that repeat in a pattern are called *periodic* because they repeat periodically or at intervals. Create a pattern and repeat it one full time and another partial time. _____

3. Trade your paper with another student. While your partner solves your pattern by completing it, you solve the one he or she gave you. Have your partner write his or her answer here.

_____

4. Use the monthly calendar grid provided by your teacher. Pick any month. Number and label the days of the week. Label the rows with *Week 1*, *Week 2*, etc.

5. What happens every seventh day? _____

6. During the third week on your calendar, what is the date of Wednesday? _____

7. What other day is most similar to your answer in Number 6? _____

8. Write in the boxes of one column some common activities that are unique to that day. For example, if you usually attend church on Sunday, write *church* on each Sunday. Continue with other events like sports practices, family nights, school events, favorite TV shows, etc.

9. Lightly color the columns, using a different color for each one.

10. How is this calendar periodic? Give several other examples of periodicity.

_____

_____

_____

Name _____

World Grid Map

This world grid map separates the earth into sections with alphanumeric coordinates. Locate the section of each of the following features and write it on the line provided. Remember to write the letter first, followed by the number, to designate each coordinate.

1. Panama Canal _____

2. England _____

3. Cape Town, South Africa _____

4. Key West, Florida, U.S.A. _____

5. Sydney, Australia _____

6. Beijing, China _____

7. Your favorite place on Earth _____

Name _____

Attraction and Distance Investigation

Follow the directions and use the provided materials to answer the following main question:

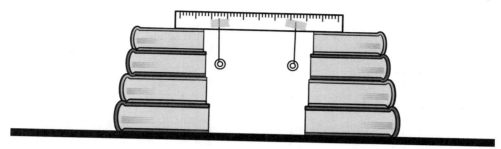

Question: What will happen when two strings with magnets on the end of each are moved closer together or farther apart?

Predict: _____

_____

Try It Out:

1. Using books, make two identical stacks about 30 cm (1 ft) high. Lay a ruler across the top of the two stacks of books.

2. Cut two pieces of string about 5 cm (2 in.) long. Tape each string about 10 cm (4 in.) apart on the ruler. To the other end of both strings, tape or tie on a strong magnet.

3. Allow the magnets to hang below the ruler. The magnets should not touch the surface below. They should be free to swing.

4. Record the distance between the top ends of the strings in the *Data Table* on **Science Notebook 5.3B Attraction and Distance Investigation, continued**.

5. Measure the angle between the string and the 0° line of the protractor. Keep the strings from swinging while measuring. Record the measurements in the *Data Table*.

6. Repeat the steps above, either increasing or decreasing the distance between the strings on the ruler, until five distances have been measured and recorded.

0° line

Name _____

Attraction and Distance Investigation, *continued*

DATA TABLE		
Trial	**Distance (cm)**	**Angle (°)**
1	10	
2		
3		
4		
5		

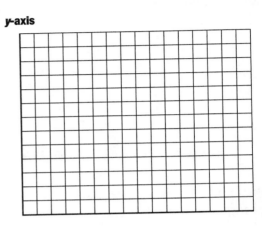

y-axis

x-axis

Analyze and Conclude:

1. Draw a graph above with the distance between strings on the *x*-axis and the angle between the string and the 0° line on the *y*-axis. Draw a line or curve that best fits the data.

2. As the distance between the strings on the ruler decreased, what happened to the angle

   between the string and the 0° line of the protractor? _____

   _____

3. Does this mean that the force of attraction between the magnets has increased or decreased?

   (Circle one.)   increased   decreased

4. This means that as the distance between magnets decreased, the force of attraction

   _____.

5. What happened when the opposite was done—when the distance between the strings was

   increased? _____

6. This means that as the distance between magnets increased, the force of attraction

   _____.

7. Compare these results to your prediction. Revise your prediction if necessary.

   _____

8. This relationship between distance and force of attraction is true not only for magnets, but also in gravity and positive/negative charges. What happens to the force of attraction between a negative electron and a positive proton when the distance between them increases?

   _____

Name _____

Chemical Families and Structure

Use the periodic table and your textbook to fill in the following:

Symbol	Atomic #	Atomic mass	Family or Group #	# of Protons	# of Neutrons	# of Electrons	# of Electrons in First Level	# of Electrons in Second Level	# of Electrons in Third Level
H									
Li									
Na									
He									
		11		10	10				
Ar	5								
Al		32	16		8	8			6
Be	4								
	12				12				
	7			15					
					16				5

Name _____

Chemical Families and Structure, continued

Use the periodic table and the chart on **Science Notebook 5.4A Chemical Families and Structure** to answer the following:

1. List the symbols of the elements on the chart in Group 1. _____

2. How many valence electrons does each of these elements have? _____

3. Give the symbol of another element on the periodic table that also has this many

   valence electrons. _____

4. List the symbols of the elements having valence electrons in the third energy level.

   _____

5. Which period of the periodic table are these elements in? _____

6. List the symbols of the nonmetals included on the chart. Remember, nonmetals generally

   have five, six, or seven valence electrons. _____

7. Repeat Number 6, but this time list the metals included on the chart.

   _____

8. List the symbols of the elements on the periodic table having four valence electrons.

   _____

9. What is true about the properties of the elements listed in Number 8?

   _____

10. What vocabulary term is given to the elements listed in Number 8? _____

11. Which period of the periodic table only has two elements in it? _____

    Why are there only two elements in this period? (Hint: Use the atomic structure and electron

    energy levels in your answer.) _____

    _____

    _____

12. Give the atomic number of another element on the periodic table with the same number of

    valence electrons as nitrogen and phosphorus. _____

Name _____

Electron Dot Diagrams

Below is a portion of the periodic table. It includes Family Groups 1–2 and 13–18 and Periods 1–6. The electron dot diagrams for a few elements have been included. Complete the chart by referring to the periodic table at the beginning of this chapter in your textbook. Use the following rules:

- Dots represent valence electrons.
- Dots can go in four places: top, bottom, left, and right of the symbol.
- Except for helium, He, spread dots out before you double them up.
- Do not put dots in the corners of the symbols.
- Do not let electron dots get confused with the dot on the letter *i*.

Periodic Table of Elements

Group #1	Group #2		Group #13	Group #14	Group #15	Group #16	Group #17	Group #18
Ḣ (Period #1)								
Li (Period #2)	Be·		Ḃ·					
(Period #3)								
(Period #4)								
(Period #5)								
(Period #6)								

Name _____

Predict the Formula

Use your answers from **Science Notebook 5.5A Electron Dot Diagrams** and the following example to predict the correct chemical formulas for the molecules formed from the given combinations of elements. (*Hint: It is possible to move some dots around so that an atom can get eight electrons. One exception to this is hydrogen. It always should get a total of two electrons, not eight.*)

Example:
sodium Ṅa and chlorine ·Ċl: combine to form sodium chloride Na:Ċl:

Valence electrons between two atoms are shared by both atoms. In this example, sodium has one valence electron that it can share. Chlorine has seven valence electrons and can accept an additional one. When the two atoms combine, both have eight electrons.

1. hydrogen and phosphorus

2. hydrogen and sulfur

3. hydrogen and fluorine

4. carbon and chlorine

5. silicon and iodine

6. nitrogen and fluorine

7. oxygen and chlorine

Name _____

Periodic Properties Graph

After reading **BLM 5.6A Periodic Properties**, graph the radii on the grid below. Use the atomic number for the *x*-coordinate and the radius as the *y*-coordinate. To do this, look at the high and low values for each axis. Decide what intervals to use. Plot each point and then connect the dots with straight lines. Label the axes and provide a title for your graph.

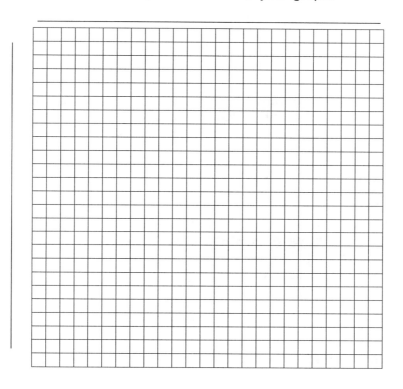

1. Before graphing, what patterns do you see in the atomic radii? _____

2. After graphing, what patterns do you see? _____

3. What is the advantage of graphing data? _____

4. Of the first 20 elements, which has the largest radius? _____

5. Which has the smallest radius? _____

6. What would you predict to be the radius of scandium (atomic number 21)? _____

7. What would you predict to be the radius of rubidium (atomic number 37)? _____

8. Atomic radius is called a *periodic property*. Using your graph, explain why.

_____

_____

_____

Name _____

Shapes of Molecules

Linear Shape

When two identical atoms bond with one central atom, like in SiO_2, they often take the shape of a line. Model this by tying or taping together two small inflated balloons.

    1. Draw a picture of how the balloons look.

    2. Using the symbols Si and O, draw an electron dot diagram of a SiO_2 molecule with a linear shape. The silicon atom is the central atom. Refer to your textbook, if needed.

Triangular Shape

When three identical atoms bond with one central atom, like in $AlCl_3$, they often take the shape of a triangle. Model this by tying or taping together three small inflated balloons.

    3. Draw a picture of how the balloons look.

    4. Using the symbols Al and Cl, draw an electron dot diagram of an $AlCl_3$ molecule with a triangular shape.

Tetrahedral Shape

When four identical atoms bond with one central atom, like in CH_4, they often take the shape of a tetrahedron. Model this by tying or taping together four small inflated balloons.

    5. Draw a picture of how the balloons look.

    6. Using the symbols C and H, draw an electron dot diagram of a CH_4 molecule with a tetrahedral shape.

Name _____

Unknown Elements

Imagine that an atom of an element could talk. Below are statements made by twenty-one different elements. Place the name and symbol of each element after the hints.

A. I have 2 valence electrons. I am heavier than zinc and lighter than silver. _____

B. I have 12 neutrons. My atomic mass is 24. _____

C. I am shiny. I have one dot in an electron dot diagram. I am lighter than carbon. _____

D. I am the "Element of Life." I like to bond with four chlorine atoms. _____

E. One of my numbers is 19. I will not conduct electricity. _____

F. I combine with one chlorine atom to make table salt. I have 12 neutrons in my nucleus. _____

G. I am the lightest element. Four of my atoms bond with one carbon atom. _____

H. I am the most common element in air. I will combine with three H atoms to form ammonia. _____

I. I combine with two oxygen atoms to make sand. I have four valence electrons. _____

J. I am in the same chemical family as nitrogen. I am lighter than argon. _____

K. One of my numbers is 16. I combine with two hydrogen atoms to make gas that smells like

rotten eggs. _____

L. I am very unreactive because I have eight valence electrons. I have 18 protons. _____

M. My properties are very similar to those of sodium. I am lighter than the element that has an

atomic number one less than mine. _____

N. You need me in your bones to make them strong. I am hard. One of my numbers is 40. _____

O. Without me, nothing would burn. One of my numbers is 16. _____

P. I have eight valence electrons. I am famous for the bright lights on Broadway. _____

Q. I am used in blimps; I have no electrons in the second energy level. _____

R. I have 18 neutrons; I will combine with one hydrogen atom to make an acid in the stomach.

_____

S. I have seven valence electrons; I have 74 neutrons. _____

T. I cannot decide—sometimes I want to be a metal, other times I want to be a nonmetal. I am a

real lightweight. Some say that I am not very interesting. _____

U. I am used as a foil around food. One of my numbers is 27. I have no electrons in the fourth

period. _____

Name _____

Sitting at the Periodic Table

Your teacher will assign you one of the elements from **Science Notebook 5.6C Unknown Elements**. After completing the first four steps below, work with your classmates to arrange the desks in your classroom into the proper arrangement of the periodic table. Sit at the appropriate desk so that your element is positioned correctly compared to the other elements in the table. Complete the remaining steps.

1. Write clues about your element from Science Notebook 5.6C: _____

_____

2. Determine the name of your element: _____

3. Explain how you decided upon the identity of your element.

_____

4. Draw a picture of an overhead view of the classroom's current desk arrangement.

5. Draw another overhead picture of the arrangement of the desks showing relationships between the elements. Label this chart with group numbers, period numbers, and element names.

Name _____

Metal Reactivity

Use the periodic table in your textbook to answer the following questions.

1. List the chemical symbols of the elements in Group 2 on the chart below.

2. For each element listed, write the number of valence electrons and the energy level number for the valence level. (Hint: There are seven energy levels. They are numbered 1–7.)

Element symbol	# of valence electrons	Energy level # for valence level

3. Most elements need eight valence electrons to be most stable. Predict what Group 2 elements will do during a chemical change. _____

_____

_____

_____

4. Explain why Group 2 elements are called a *chemical family*. _____

_____

_____

5. Which Group 2 element will react most easily? _____

6. Which Group 2 element is the hardest to make react? _____

7. Use atomic structure to explain the properties described in Questions 5 and 6.

_____

_____

8. Which metal is more reactive in Group 13—aluminum or thallium?

_____

Name _____

Nonmetal Reactivity

Use the periodic table in your textbook to answer the following questions.

1. List the chemical symbols of the elements in Group 16 on the chart below.

2. For each element listed, write the number of valence electrons and the energy level number for the valence level. (Hint: There are seven energy levels. They are numbered 1–7.)

Element symbol	# of valence electrons	Energy level # for valence level

3. Most elements need eight valence electrons to be most stable. Predict what Group 16 elements will do during a chemical change. _____

_____

4. Which Group 16 element will react most easily? _____

5. Which Group 16 element is the hardest to make react? _____

6. Use atomic structure to explain the properties described in Questions 4 and 5.

_____

_____

7. Which nonmetal is more reactive in Group 15—nitrogen or phosphorus? _____

8. Why are Group 18 elements nonreactive? _____

_____

_____

9. Describe how the structure of the atom causes its properties.

_____

_____

_____

Name _____

Vocabulary Review

Write the number of the word on the blank of its definition.

1. atomic number

2. strong force

3. chemical family

4. electron dot diagram

5. metal

6. periodic table

7. period

8. electrostatic charge

9. nonmetal

10. atomic mass

11. valence electron

12. electron energy level

_____ a property of matter that can be either positive, negative, or neutral

_____ an element that usually contains one, two, or three valence electrons and that is typically a shiny, hard, good conductor

_____ a symbol in which dots represent valence electrons

_____ the location of electrons surrounding the nucleus in the atom

_____ the attractive force that holds the nucleus together

_____ a group of elements that have similar properties

_____ the number of protons plus the number of neutrons in one atom

_____ the number of protons in one atom of a particular element

_____ an electron in the outermost energy level of an atom

_____ a chart that arranges elements having similar properties into the same group

_____ a horizontal row of the periodic table

_____ an element that usually contains five to seven valence electrons and that is typically a dull, brittle, poor conductor

Name _____

Chapter 5 Review

Use the periodic table to complete the exercises below.

1. Name an element that has seven valence electrons. _____

2. Name an element similar to the element with an atomic number of 12. _____

3. $^{34}_{79}$Se has an atomic mass of _____ and _____ neutrons in one atom.

4. An atom of sodium (atomic number of 11) has _____ electrons in the first energy level,
   _____ electrons in the second energy level, and _____ electron(s) in the third energy level.

5. The force of _____ between positively charged protons is overcome by the _____
   force made possible by the presence of neutrons in the nucleus.

6. $^{9}_{19}$F has _____ protons, _____ neutrons, and _____ electrons in one atom.

7. Molecules take certain shapes, such as lines or triangles, because of the force of _____
   between negatively charged _____.

8. _____ electrons are represented by dots in _____ _____ diagrams.

9. Nitrogen (atomic number of 7) is in Group _____ and Period _____.

10. How many hydrogen atoms are needed to form a molecule with one atom of phosphorus
    (atomic number of 15)? _____ Use electron dot diagrams to show how this occurs.

    _____

11. Draw the electron dot diagram for an atom of oxygen. _____

12. Classify elements with the following atomic numbers as being either metals or nonmetals: 4,
    9, 11, and 16. _____

13. Think of the phrase *Which came first, the chicken or the egg?* Now consider, answer, and
    explain *Which came first, structure or properties?* _____

    _____

    _____

    _____

    _____

    _____

Name _____

Demonstrating States of Matter

Read and complete the exercises below.

1. On the following chart, list at least three properties of each of the three states of matter.

Solids	Liquids	Gases
1.	**1.**	**1.**
2.	**2.**	**2.**
3.	**3.**	**3.**

2. With your group, design an imaginative way to demonstrate each of the three states of matter. You may use music, art, drama, computers, or any other method that your teacher approves.

   a. Choose one property from each state of matter to demonstrate.

     solid: _____ liquid: _____ gas: _____

   b. Describe how you plan to model each of the three states. Include the special way in which your model illustrates the particular property you have chosen for that state of matter.

     solid: _____

     _____

     _____

     liquid: _____

     _____

     _____

     gas: _____

     _____

     _____

   c. List the materials you will need for each model.

     solid: _____

     liquid: _____

     gas: _____

Name _____

Evaluation of Models

Perform your demonstration for the class. Carefully observe the other groups while they perform their demonstrations. Evaluate each group as described below. Number each group according to the sequence in which the demonstration was given. For example, the first group to give their demonstration will be Group 1. Follow your teacher's instructions for how he or she would like you to fill in the chart regarding your own group number.

1. List the property that was modeled by each group for each state of matter.

	Solid	Liquid	Gas
Group 1			
Group 2			
Group 3			
Group 4			

2. Describe how each property was illustrated.

	Solid	Liquid	Gas
Group 1			
Group 2			
Group 3			
Group 4			

3. Evaluate each model. Did it accurately depict the state? Was it clear and easy to understand?

	Solid	Liquid	Gas
Group 1			
Group 2			
Group 3			
Group 4			

Name _____

Heat Capacity of a Water Balloon

Watch your teacher perform two experiments and then answer the following questions:

1. Describe what your teacher did. _____
   _____

2. Why did the first balloon burst? _____

3. Predict what will happen to the second balloon. _____
   _____

4. What happened to the second balloon? _____
   _____

5. Explain the reason for the result in *Question 4.* _____
   _____

6. Where does the heat energy from the flame go in each situation? _____
   _____
   _____
   _____

7. Are there any markings on the water-filled balloon? _____

8. Has the water changed? If so, how? _____
   _____

9. What property of water is demonstrated in this activity? _____

10. Compare the heat capacity of water to the heat capacity of air based on this test.
    _____
    _____
    _____
    _____
    _____
    _____

Name _____

Heating Curve of Water

Use the graph on **BLM 6.3A Heating Curve of Water** to answer the following questions about water, its properties, and its heating curve.

1. What state(s) is (are) present at 1 min 30 s? At 22 min? How do you know? _____

_____

_____

2. What do the flat regions of the graph represent? _____

3. If heat is constantly being added to the water, why are there regions of the graph where the

temperature is not increasing? _____

_____

_____

4. What is the melting temperature represented on the graph? _____

5. What is the boiling temperature represented on the graph? _____

6. Consider the water balloon trial. Use the term *heat capacity* and refer to the heating curve of

water to explain what kept the balloon from bursting. _____

_____

_____

7. On the graph, clearly label the two main states represented.

8. On the graph, clearly label the state changes of melting and boiling.

9. Imagine that instead of heating ice, the experimenter had cooled water vapor. Describe how the graph would be different and then sketch an example in the space provided.

_____

```
110 ─
100 ─
 90 ─
 80 ─
 70 ─
 60 ─
 50 ─
 40 ─
 30 ─
 20 ─
 10 ─
  0 ─
-10 ─
```

10. On your sketch, clearly label the *condensation point* and the *freezing point*.

11. On your sketch, clearly label each state—*ice*, *liquid water*, and *water vapor*.

Name _____

Penny Puddles

Follow the directions below.

1. You will receive a penny, an eyedropper, a paper towel, and a small cup from your teacher. Fill the cup with water. Place drops of water on the "heads" surface of the penny, using the eyedropper. Count how many drops you add. On the chart below, record the number of drops it takes before the water spills over the edge of the penny. Dry the penny. Repeat the trial two more times, and record the results.

Surface	Trial 1	Trial 2	Trial 3
front (heads)			
back (tails)			

2. Next, turn the penny over and perform three trials on the "tails" side. Complete the chart.

3. Was there a difference in the number of drops that were required in any of your trials? _____

4. What caused the differences? _____

_____

_____

5. What was it that caused the water to stay on the penny? _____

6. What kind of bonds broke when the water spilled over the penny? _____

7. With your partner, join with three other pairs of students and create a scatter plot to show the number of drops placed on the heads side in each of your three trials. (You should have a total of 12 data points.)

_____

8. Explain what was happening with water throughout this activity. Use the terms *polarity*, *hydrogen bonds*, and *surface tension* in your response.

_____

_____

_____

_____

Name _____

Solubility Ability

In your group, predict the solubility of the substances in the chart. After making your predictions, follow the directions to test which substances actually are soluble in water. You may not have all the substances in the chart, so be prepared to hear the results of other groups and to share your own with the class. Place an *X* on the chart to the left of the substances that your group is testing.

Place each substance you are given into 100–250 mL of water. Wait for 5 seconds and then stir the solution for 10 seconds. Observe your results and write your answer in the appropriate box. For items that do not come in a predetermined size, use 5 mL (1 tsp).

Substance	Solubility Prediction (yes or no)	Actually Soluble (yes or no)
calcium		
vitamin B (crushed)		
Smarties® (crushed)		
sugar cube (crushed)		
salt		
pepper		
vegetable oil		
rubbing alcohol		
hydrogen peroxide		
chocolate		
toothpaste		
white school glue		

1. Scientifically, if an item only partially dissolves, is it soluble or insoluble in water?

_____

2. Explain why you think you predicted incorrectly for any substances that the prediction and the

actual solubility do not agree. _____

_____

_____

_____

3. Are there any similarities between the items that would not dissolve?

_____

_____

Name _____

pH Scale

Follow the steps below to test the pH of some common household items.

1. Acquire each item from your teacher along with an appropriate number of pH test strips. Keep items separated.
2. Complete *Step 1* on the next page.
3. Slowly dip a pH test strip into each solution and compare the strip with your teacher's transparency to determine the pH. For fruit items, gently squeeze the piece you are using to produce a small amount of juice and touch the strip to the juice.
4. Once you have determined the pH, record it on the chart below in the *pH* column.
5. When all of the items have been tested and the pH of each is determined, write in each item in the appropriate place on the pH Scale. You may need to make marks in-between the provided markings.
6. On the right of the pH Scale, label the two sections as either *acids* or *bases*.
7. Once those items have been listed, write in each item from the text in your science textbook and answer the questions that follow.

Item	pH
ammonia	
apple	
baking soda	
banana	
black coffee	
liquid dish soap	
egg	
lemon juice	
milk	
soda water	
tomato juice	
toothpaste	
vinegar	
tap water	
window cleaner	

pH Scale

14

0

Name _____

pH Scale, continued

Answer the following questions, beginning with *Question 2*:

1. Predict which items will be acids and which will be bases.

   acids: _____

   bases: _____

2. Circle items in *Question 1* that you predicted correctly.

3. Which is the strongest base tested? How do you know? _____

   _____

   _____

4. Which is the weakest acid? How do you know? _____

   _____

   _____

5. Are there any weak bases or strong acids? Which are they and how are they identified?

   _____

   _____

   _____

   _____

6. Hydrochloric acid has a pH of 0. Bleach and oven cleaners have pH levels of 12 and 13. Use this information to describe why humans should not ingest strong bases and acids.

   _____

   _____

   _____

7. What is the term for substances that prevent drastic changes in pH? _____

8. When do scientists use the substances in *Question 7*? _____

   _____

   _____

   _____

   _____

Name _____

Indoor Rain

Watch your teacher perform the lab demonstration and answer the following questions:

1. What was the temperature of the water when the test began? _____

2. At what temperature did the water boil? _____

3. What had to be added to get the water to boil? _____

4. Describe what happened after the water began to boil. _____

   _____

   _____

5. What was the purpose of the cookie sheet? _____

   _____

6. Was all of the water vapor converted back to liquid water? Why? _____

   _____

   _____

7. What is the process in *Question 6* known as? _____

8. What happened to the ice? _____

9. Why did this happen to the ice? _____

   _____

10. Watch as your teacher draws a cooling curve and a heating curve on the board. Describe the curves and what happens at the melting and boiling points.

   _____

   _____

   _____

   _____

   _____

   _____

Name _____

Temperature Conversions

Use the equations listed below to convert from one temperature scale to another. Some problems may require using two equations or rearranging the equations. Show your work.

$$K = C + 273$$
$$C = \tfrac{5}{9} \times (F - 32)$$
$$F = (\tfrac{9}{5} \times C) + 32$$

1. Convert 61° C into K.

2. Convert 68° F into °C.

3. Convert 200° F into K.

4. Convert 100° C into °F.

5. Convert 215 K into °C.

6. Convert 300 K into °F.

7. Convert 0° F into °C.

8. Convert 150 K into °F.

9. Convert –40° F into °C.

10. Convert –459.4° F into K.

11. Convert 3 K into °C.

12. Convert 0° C into °F.

Name _____

Heat Insulation

Determine which material maintains the temperature of a liquid the longest. Follow the steps below to answer the questions:

1. Acquire one glass, ceramic, and foam container from your teacher. Fill each container to within 1 cm ($\frac{1}{2}$ in.) from the lip with the hot water your teacher has prepared.
2. Immediately cover each container with aluminum foil. Be sure to leave a small opening in the foil at the edge of each container to insert a thermometer.
3. After one minute, take the temperature of each container's contents. Record the temperatures in the chart below, making sure to designate which scale you are using (Celsius or Fahrenheit).
4. Wait 1 minute. Record the new temperature in the same order as you did in *Step 3*. Be sure that each container remains covered.
5. Repeat *Step 4* until the chart is completed. Answer the questions and wait for your teacher's instructions.

Time (in minutes)	Temperature in glass container	Temperature in ceramic container	Temperature in foam container
0			
1			
2			
3			
4			
5			
6			
7			
8			
9			
10			

1. Which substance does the best job insulating in this experiment? _____

2. Which substance does the poorest job insulating in this experiment? _____

3. How are the best and the worst insulators different in this test? _____
_____

4. What role does insulation play in a building and which kind of material would make the best insulator for a building based on this test? _____
_____

5. Do you think that the results using ice would be different concerning each material's insulating ability? Explain. _____
_____
_____

Name _____

Heat Work

Perform the test below to see how heat can accomplish work. When you have finished, clean up your area and answer the questions.

1. Tie a 61 cm (2 ft) long piece of string around the neck of a clean, dry, glass bottle. Tie a second 61 cm (2 ft) long piece of string about 2 cm (1 in.) above the base.
2. Place a meterstick on a table or desk so that 45 cm (18 in.) extend off of the edge. Place several books on top of the portion of the meterstick on the desk to create stability.
3. Tie both strings to the meterstick so that the bottle is suspended horizontally and parallel to the floor.
4. Place two wooden match heads into the bottle and position them near the center.
5. Gently insert a rubber stopper into the mouth of the bottle far enough to ensure closure. Refer to the image to ensure appropriate setup.
6. After comparing your apparatus to the image, light a candle and carefully hold the flame directly below the match heads.

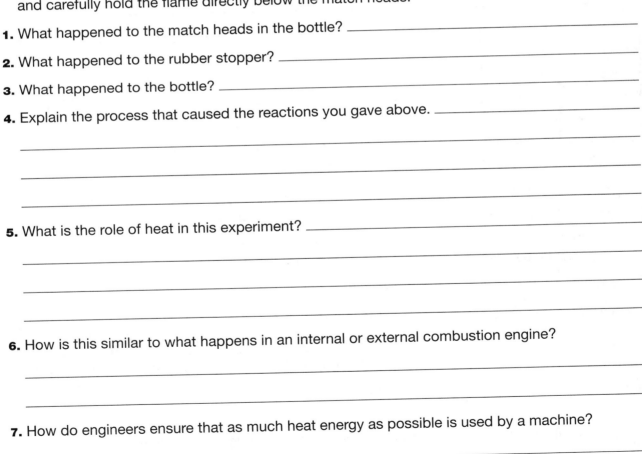

1. What happened to the match heads in the bottle? _____

2. What happened to the rubber stopper? _____

3. What happened to the bottle? _____

4. Explain the process that caused the reactions you gave above. _____

_____

_____

_____

5. What is the role of heat in this experiment? _____

_____

_____

_____

6. How is this similar to what happens in an internal or external combustion engine?

_____

_____

7. How do engineers ensure that as much heat energy as possible is used by a machine?

_____

_____

Name _____

Vocabulary Review

Write the letter of the correct definition in the blank for each corresponding vocabulary term.

_____ **1.** sublimation

a. a temperature scale that begins at absolute zero

_____ **2.** plasma

b. a substance with a pH less than 7.0

_____ **3.** ion

c. the elastic property of liquids that allows drops to form and creates the smallest area possible at the surface of a liquid

_____ **4.** Kelvin scale

d. the ability to store and release heat energy

_____ **5.** absolute zero

e. the change from a solid to a gas without passing through the liquid state

_____ **6.** heat capacity

f. a range of values from 0 to 14 that indicates the concentration of hydrogen ions in a solution

_____ **7.** polarity

g. a superheated gas composed of electrically charged particles

_____ **8.** hydrogen bond

h. the quality of having opposite charges for different points of a molecule

_____ **9.** surface tension

i. a substance with a pH greater than 7.0

_____ **10.** solubility

j. a weak bond that connects hydrogen-containing molecules at polar areas

_____ **11.** viscosity

k. an atom that has gained or lost one or more electrons

_____ **12.** acid

l. the resistance of a liquid to flow

_____ **13.** base

m. the lowest possible temperature at which matter has no remaining thermal or heat energy

_____ **14.** pH scale

n. the change from a gas to a solid without passing through the liquid state

_____ **15.** deposition

o. the ability of one substance to dissolve in another

Name _____

Chapter 6 Review

Answer the following questions without using your textbook.

1. List four states of matter and one example of each. _____

_____

2. As a substance heats up and cools down, it goes through state changes. Use water as an example, and list the phase changes necessary for water as it changes states of matter.

_____

_____

3. Which state of matter has more energy than the gas state? _____

4. What was a problem with early attempts to measure temperature? _____

_____

5. Who was the first to solve these temperature problems? How? _____

_____

6. What is unique about an absolute scale? _____

_____

7. List three properties or abilities of water that are due to its polarity. _____

_____

8. Circle the correct answer. When a fluid flows slowly it is said to have a

   (high viscosity / low viscosity).

9. List two examples of acids and identify whether they are strong or weak. _____

_____

10. List two examples of bases and identify whether they are strong or weak. _____

_____

11. The concentration of which atom determines pH on the pH scale? _____

12. What are two styles of engines studied in this chapter and how are they different?

_____

_____

13. What is the cycle for the refrigerant in a cooling system? _____

_____

_____

Name _____

Changes in State

Ice melts. Water freezes or boils. Steam condenses. Each of these is a change in the state of matter.

1. List the three states of matter that are most common on Earth.

_____ _____ _____

2. In what state of matter is the cold ice cream? _____

3. In what state of matter is the chocolate topping? _____

Your teacher will place a scoop of ice cream in a cone, and top it with chocolate. After eating your treat, answer the following questions:

4. What happened to the chocolate topping? _____

5. If there was chocolate on the cone, in what state of matter was it? _____

6. In what state of matter was the chocolate on the ice cream? _____

7. Use the terms *heat energy* and *temperature* to explain why this happened. _____

_____

8. What would happen to the chocolate if it were taken off the ice cream? _____

9. Matter that flows is called a *fluid*. Identify a fluid in this activity. _____

10. Identify three things that you ate in this activity that were not fluids.

_____ _____ _____

11. What state of matter is not a fluid? _____

12. Identify another state of matter that will flow besides liquids. _____

13. Fill in the chart below. The first line has been completed for you to show that evaporation changes a liquid to a gas.

CHANGE IN STATE	CHANGES A ?	TO A ?
Evaporation	liquid	gas
Condensation		
Freezing		
Melting		
Sublimation		
Deposition		

Name _____

Demonstrating States of Matter

Scientific models represent God's creation in ways that help us understand and visualize the way the world is made. Three states of matter can be modeled in the following way:

State of Matter	Model of Molecules/Atoms
solid	a military unit in close formation, standing at attention
liquid	a group of teenagers, walking down a sidewalk and talking
gas	children playing the game of tag

As a group, demonstrate the way molecules and atoms act in each state of matter as described above by having each person act like an atom or molecule. Then answer the following questions.

1. Do the molecules and atoms move in all three states of matter? _____

2. In which state of matter do the particles move the fastest? _____

3. In which state of matter are the particles the closest to each other? _____

4. In which state of matter are the particles the farthest apart? _____

5. Which state(s) of matter are able to flow? _____

6. Which state(s) of matter cannot be compressed? _____

7. Which state of matter is the least dense? _____

8. As a group, design a model of the states of matter using animals that can be found in a zoo.

Describe your model. _____

_____

_____

_____

9. Design and describe another states-of-matter model, using something totally different.

_____

_____

_____

Name _____

Archimedes' Principle

Archimedes' Principle explains why various objects sink, float, or rise. Regardless of whether the fluid is a liquid or gas, if the mass of the fluid displaced equals the mass of the object, the object will float. If the masses are equal, the weights will also be equal. The upward force caused by the displaced fluid is called *buoyancy*.

Follow the directions and answer the questions below:

1. Determine the volume of a block of wood in the following way:

    a. Measure in centimeters and record the length, width, and height of the block.

    length = _____   width = _____   height = _____

    b. Multiply these three dimensions to determine the volume in cubic centimeters.

    volume = _____

2. Measure the mass of the block of wood in grams.

    mass = _____

3. If the block of wood were totally submerged in water, what would the volume of the displaced water be? _____

4. Density of water is 1 gram per cm^3. Determine the mass of the displaced water in Step 3.

_____

5. Which is greater, the mass of the block or the mass of the displaced water?

_____

6. What will the block of wood do when completely submerged in water?

_____ Try it!

7. When the block of wood floats, what volume of water is displaced? _____ Explain.

_____

8. Repeat Steps 1–6 with a piece of metal or rock. If the metal or rock has an irregular shape, measure its volume by the displacement of water. Record the answers below for each step.

    length = _____   width = _____   height = _____

    volume of metal or rock = _____ volume of displaced water = _____

    mass of metal or rock = _____ mass of displaced water = _____

    Which has the greater mass—the water or metal (rock)? Circle your answer.

Name _____

Pascal's Principle

How would you like to make your partner float on a pillow of air? Using Pascal's Principle and air pressure, you can do it as easily as inflating a balloon.

1. Close an empty 3.75 L (1 gal) plastic bag and cut a hole at the bottom corner, just large enough to hold the tubing discussed in Step 2.

2. Place the end of a 30 cm (1 ft) length of plastic or rubber tubing through the opening in the plastic bag. Tape around the opening so that the tube is secure and no air can enter or leave the bag except through the tube.

3. Practice by placing a book on the bag. Blow several times into the tube to inflate the bag just

   like a balloon. What happens? _____

4. Now set the empty bag in a chair. Have your partner sit on the bag. Use the tubing to inflate

   the bag. Record what happens. _____

5. For health reasons, no one else should use your tubing. Your partner should now use his or her bag and tubing to try to make you float on air.

6. Write Pascal's Principle.

   _____

   _____

7. Use Pascal's Principle to explain how this activity works.

   _____

   _____

   _____

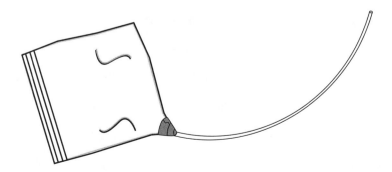

Name _____

Pressure Experiments

1. Perform *Experiment 1* described on **BLM 7.4B Directions: Pressure Experiments,** using two table tennis balls, string, tape, and a straw. Describe what happens when you blow between the table tennis balls.

_____

2. Try blowing harder and softer. Which offers more results? _____

Explain why. _____

3. Increase and decrease the distance between the balls. Which offers more results? _____

_____

Explain why. _____

4. Perform *Experiment 2* from BLM 7.4B and describe what happens, explaining why

this happens. _____

_____

_____

5. Perform *Experiment 3* described on BLM 7.4B. Describe what happens and why. _____

_____

_____

6. Perform *Experiment 4* from BLM 7.4B and describe what happens. _____

_____

7. Try bending the short part of the straw about 20 degrees away from vertical. Can you keep the

table tennis ball in the stream of air now? _____ Explain why. _____

_____

_____

8. Perform *Experiment 5* from BLM 7.4B. What happens? _____

Explain why this happens. _____

_____

9. Perform *Experiment 6* from BLM 7.4B. What happened when you let go of the card?

_____

Explain why this happens. _____

_____

Name _____

Bernoulli Bags

With a partner, follow the steps given below, record your results, and answer the questions that follow regarding Bernoulli's Principle.

1. You and your partner each get a small trash bag from your teacher.

2. In your assigned area, make sure that all of the air is out of one of the bags. One of you hold the bottom of a bag and encourage the other person as he or she blows it up like a balloon.

How many breaths did it take? _____
Switch roles and let the second person blow up his or her bag while the first person holds the

bottom of the bag. How many breaths did he or she take? _____

3. Switch roles again. Make sure there is no air in the bag. The person blowing should open the bag about 30 cm (1 ft) at the opening and place his or her face as close as possible to the opening. Blow into the bag several times. The bag does not get any bigger after the first breath! Let the second person try it.

4. Have the first person squeeze all the air out of their bag and hold the mouth of the bag open 30 cm (1 ft) once more. This time hold the bag about 30 cm (1 ft) away from your mouth and blow straight into the opening. Close the bag when you stop blowing.

What happened? _____

Have the second partner try this step as well. Did it happen again? _____

5. Why was the bag so difficult to inflate the first time? _____

_____

6. Why was the bag so difficult to inflate the second time? _____

_____

7. Why did the bag inflate with the last attempt? Explain, using Bernoulli's Principle.

_____

_____

8. Explain why the bag inflated in the last attempt without referencing Bernoulli's Principle.

_____

_____

9. In the last trial, did you expect the bag to inflate well or poorly? Why?

_____

_____

Name _____

Paper Airplanes

Complete *Steps 1–10* on **BLM 7.5B Directions: Paper Airplane** to make a slow-flying paper airplane that produces lift easily. Then follow the steps below.

1. Fly your airplane five times, making adjustments between each flight. Each time, use your stopwatch to determine how many seconds your airplane stays in flight. Record the longest time. What did you discover that helped the airplane stay in the air longer?

   _____

   _____

2. Repeat *Step 1*, but this time determine the greatest distance you can fly your airplane. Record the number of steps from your starting point to the landing point for your longest

   flight. _____ What helps the airplane fly greater distances? _____

   _____

3. Complete *Step 11* from BLM 7.5B. Bend the elevators in such a way that both elevators bend in the same direction. Make your airplane fly. How do the elevators affect the flight?

   _____

4. Bend the elevators the opposite way from *Step 3*. How does this affect the flight?

   _____

5. Bend the elevators so that they point in opposite directions. Predict how this will affect

   the flight. _____ Test your prediction. Was your prediction correct?

   _____ Explain what happened. _____

6. Make another adjustment to your airplane. Describe your adjustment and how you predict it

   will affect the flight. _____

7. Test your adjustment and record your observations. _____

   _____

8. What provides the thrust for your paper airplane? _____

9. What causes the drag for your airplane? _____

10. In terms of weight, what is the advantage of paper airplanes? _____

11. What is providing lift for your paper airplane? _____

    _____

Name _____

Fly a Kite

Take a kite outside on a day when there are steady breezes. Do not fly it near buildings, trees, or overhead lines. Put your back to the wind. Hold onto your spool of string and release the kite. As the kite begins to fly, slowly release more string so that the kite flies higher. Think about the four forces involved in flight as you control your kite. When your time is up, wrap your string around the spool and bring your kite to the ground.

1. What provides the thrust for a flying kite? _____

2. What is the purpose of the tail of a kite? _____

3. What produces the lift for a kite? _____

4. What happens if you pull hard on the string that is tied to a flying kite?

_____

5. Explain why this happens. _____

6. What happens if you run into the wind while flying a kite?

_____

7. Why does this happen? _____

8. What happens if you run with the wind while flying a kite?

_____

9. Explain why this happens. _____

10. Name three important things to remember in order to build a successful kite.

_____

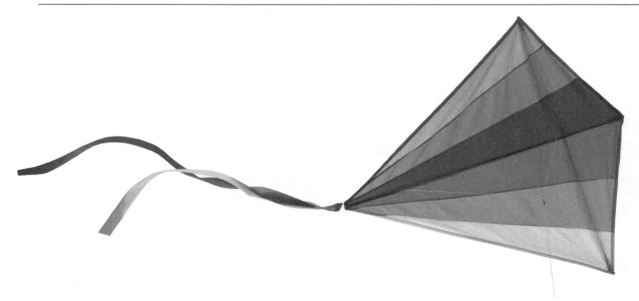

Name _____

Flying Objects

Follow the directions to investigate Bernoulli's Principle.

Question: How can Bernoulli's Principle overcome gravity?

Objects like balloons and table tennis balls are attracted by Earth's gravitational field with a force called *weight*. Yet, through the movement of fluids, the effect of gravity can be overcome. This experiment will use a hair dryer to create a miniature wind tunnel.

Predict: What will happen if a tube is placed above a stream of air that has a ball or some

other object suspended in it?_____

Try It Out:

1. Turn the hair dryer on with the heat off, if possible, and point it straight up. Using high power, place a table tennis ball in the stream of air. Describe what happens.

_____

2. Experiment with three other similar objects to discover if the result is similar. Turn the hair dryer off when not in use. List each object and how it performed.

_____

_____

3. Place the table tennis ball back in the stream of air and change to low power. What happens?

_____

4. Return to high power. Slowly tilt the hair dryer to the side while suspending the table tennis ball in the stream of air. What happens?

_____

5. Repeat *Step 1* and slowly bring a short tube into and aligned with the stream of air until the

tube is directly above the table tennis ball. Describe what happens.

_____

6. Predict what will happen when a longer tube is used. _____

Try it out. Record what happens. _____

_____

7. Predict what will happen when a tube with a larger diameter is used. _____

Try it out. What happens? _____

7.6B
NOTEBOOK

Name _____

Flying Objects, continued

Analyze and Conclude:

1. When fluids move through smaller objects like the tube, the air speeds up so that the same amount of air can move through in the same amount of time. What happens to the internal air

pressure when the air speeds up through the tube? _____

_____

2. What is this fact of science called? _____

3. Did the air move faster through the smaller-diameter tube or the larger-diameter tube?

_____

4. Which diameter tube had the lowest internal pressure? _____

5. Use the reasoning in the previous four questions to explain what happens when a tube is placed over a stream of air that has an object suspended in it.

_____

_____

_____

_____

6. Explain what happens when a longer tube is used.

_____

7. Explain what happens when the hair dryer is tilted away from vertical.

_____

_____

_____

8. Use Bernoulli's Principle to explain how gravity can be overcome.

_____

_____

Name _____

Balloon in a Bottle

1. Read this page and predict what will happen in *Steps 5* and *6*.

_____

2. Take a close look at the bottom of the plastic bottle given to you. What is in the bottom of the bottle?

_____

3. Place the balloon inside the bottle and stretch the neck of the balloon over the mouth of the bottle.

4. Place a wet finger near the hole on the bottom of the bottle. Inflate the balloon inside of the bottle by blowing into the balloon at the mouth of the bottle. Describe what happens.

_____

_____

5. When finished inflating, place your wet finger over the hole on the bottom of the bottle. Remove your mouth from the balloon. Describe what happens.

_____

_____

6. After waiting a short time, remove your finger from the hole. What happens?

_____

7. Predict what will happen if you blow into the tiny hole on the bottom.

_____

8. Try it out. What did happen? _____

_____

9. Use the concept of air pressure to explain the entire activity.

_____

_____

Name _____

Plungers and Bottles

Plungers are used all over the world. You will examine how they work and just what helps get rid of pesky clogs. Follow the steps below and answer the questions:

1. Take two clean, dry plungers and push their ends together. Record what happens.

_____

2. Now slightly wet the rims, push them together, and record the results.

_____

3. Use the term _pressure_ to explain the reasons for the results of _Steps 1_ and _2._

_____

_____

_____

In this activity, you will examine what happens when you try to blow some paper into a bottle and why those results occur.

1. Take a bottle used in **Science Notebook 7.6C Balloon in a Bottle** and place tape over the hole in the bottom.

2. Set the bottle on its side. Wad up a small piece of paper and place it in the mouth of the bottle so that it covers half of the bottle's opening.

3. Blow gently into the mouth of the bottle. What happens? _____

_____

4. Repeat _Step 3_ several times, trying different techniques. Write down two other techniques to try to blow the paper into the bottle and indicate whether or not they worked.

Technique 1: _____ Did it work? (Y / N)

Technique 2: _____ Did it work? (Y / N)

5. Describe why you think your techniques did or did not work in _Step 4._ Use the term _pressure_ in your answer.

_____

_____

6. Relate how plungers do their work to what happened with the bottle. Focus on the impact of

pressure. _____

_____

Name _____

Natural Helicopters

Maple trees spread seeds in a unique way. Each seed is contained in an airfoil-like structure that produces rotation and lift similar to a helicopter. When the seed breaks away from the tree, the wind moves the seed a greater distance because of the shape of the airfoil connected to the seed. You will model these seeds in this activity.

1. Obtain a pattern from your teacher.

2. Cut along the dotted lines. Fold the paper carefully along the solid lines. *Sections A* and *B* need to be folded in opposite directions from one another. *Sections C* and *D* need to be folded toward one another, and then fold the bottom square up.

3. Write your name on your helicopter.

4. Attach a small paper clip on the bottom so that the folded square is held in place. The paper clip and square mimic the position and weight of a maple seed and stem.

5. Hold the helicopter just under *Sections A* and *B* or at the paper clip, and toss the helicopter into the air. Repeat several times. Describe what happens. _____

_____

_____

6. Two vertical forces are involved during flight: weight and lift. Does the weight change when

the helicopter rotates? _____ Does the lift change when it rotates? _____

7. If lift is generated by the spinning helicopter, why does it not fly upward?

_____

_____

_____

8. Explain what must happen for your helicopter to move up while rotating.

_____

_____

_____

Name _____

Table Tennis Ball Curves

1. Use a marker to draw a circle at the top of a table tennis ball. Make another circle at the side of the ball, using a different colored marker. This will help you see the spin of the ball. Spin the ball and watch it rotate. Try this several times, using spins of different directions.

2. Move to an open area a safe distance from anyone else. Place the ball in the tube. Use your hand to keep the ball from falling out of the bottom of the tube. Swing the tube quickly as you would a baseball bat. Observe the flight of the ball. You may need to tilt the tube downward a little before swinging. Repeat several times until you get consistent results. Describe the path

   of the ball. _____

3. Repeat *Step 2*, but swing the tube from the opposite side of your body. In other words, if you swung the tube like a right-handed batter the first time, swing it like a left-handed batter this

   time. Describe the path of the ball. _____

4. Swing the tube over your head as in a tennis serve. Describe the path of the ball.

   _____

5. Swing the tube toward the ground as in a golf swing. Describe the path of the ball.

   _____

6. Add a piece of sandpaper to the inside of the tube's end, with the rough side facing inward. It may be necessary to glue or tape the sandpaper to the inside of the tube. Repeat *Steps 2–5* by holding the end of the tube opposite from the sandpaper and describe what happens.

   _____

7. Which type of swing(s) produces topspin? _____

8. Which type of swing(s) produces backspin? _____

9. Which type of swing(s) produces sideways spin? _____

10. Draw a picture of a ball from the top view perspective. Show how the ball is spinning and which direction the ball will move. Include another picture of the opposite type of spin.

11. Draw a picture of a ball from the side view perspective. Show topspin and the direction the ball will move. Include another picture of backspin.

Name _____

Vocabulary Review

Read the following sentences and use the Word Bank to fill in the blanks. Each term or phrase is a part of the chapter vocabulary.

Word Bank	airfoil	weight	Archimedes	Bernoulli	buoyant force	lift
	hydraulics	drag	Pascal	pressure	air pressure	thrust

1. The amount of force exerted on a certain area is the _____.

2. The force of gravity acting down on all objects, including aircraft, is called

_____.

3. The principle of fluids that explains that pressure is equally distributed in all directions is

remembered by the name of _____.

4. The force on an airplane that pushes it forward is known as _____.

5. The principle of fluids showing that pressure decreases as a fluid speeds up is credited to

_____.

6. The force on an airplane that pushes it up is the _____.

7. The shape of an airplane wing is called the _____.

8. The backward force on an airplane is known as _____.

9. The principle of fluids discovered by _____ states that a submerged

object is pushed upward by a force equal to the weight of the fluid it displaces.

10. A system of _____ uses fluids to increase or transfer forces.

11. When a ship displaces water, _____ pushes up against the ship.

12. The force of air exerted on a particular area is the _____.

Name _____

Chapter 7 Review

1. Complete the following chart by adding *yes* or *no* regarding the properties of the states of matter:

State of matter	Does it take the shape of its container?	Does it fill the volume of its container?
solid		
liquid		
gas		

2. State two factors that help determine which state of matter is present.

a. _____   **b.** _____

3. Matter that can flow is called a _____.

4. Three properties of fluids that are named after scientists are _____,

_____, and _____.

5. Which of these principles explains why hot air balloons float in air? _____

6. Which of these principles explains why hydraulic systems work? _____

7. Which of these principles explains why ships float in water? _____

8. Which of these principles help explain lift? _____

9. Label the following drawing with the appropriate force involved in flight:

_____

_____   _____

_____

10. If thrust is greater than drag, the airplane is _____.

11. Explain how Newton's Law of Action and Reaction is involved in flight.

_____

_____

12. Explain how the shape of the airfoil and the angle of the airfoil are involved in flight.

_____

_____

Name _____

Motion Notion

For each of the pictures below, describe at least two examples of motion that are taking place.

_____
_____

_____
_____

Name _____

Perceiving Motion

As your teacher reads you the story of Oliver's trip to his grandpa's house, underline examples of motion and any details about motion. When the story is finished, answer the questions below.

Oliver gazed out the car window as he was traveling to his grandpa's house. He noticed that the electrical poles seemed to whiz by the car. All of a sudden he thought to himself, "Those poles are not even moving! It is funny how they look like they are moving very fast when they are actually standing still." He thought about it for a few minutes, but was not sure how to explain it. Oliver decided to read his book.

After a while he glanced out his window again and saw a boy about his age in the backseat of the car beside him in the next lane. He was also reading a book. That car and the boy in it appeared to be moving slowly backward. That seemed odd to Oliver, so he focused his eyes on the scenery that he could see through the boy's car windows. Now he could tell that both cars were moving forward. He wondered how both cars could be moving forward, when the other car looked like it was moving backward. Thinking about it gave him a headache!

A little while later, Oliver's father drove the car onto a ferry in order to cross a bay. He parked the car and set the brake. He and Oliver walked to the back of the deck and waited for the ferry to move. Oliver could hear the large engine of the ferry rumbling and he could feel the boat vibrating beneath his feet. The ferry began to move. He looked at the gate at the end of the road leading to the ferry. It seemed to be getting farther and farther away. He looked over the back of the deck at the water behind the boat. The water was churning and flowing backward from the boat, yet the boat was moving forward.

After what seemed like a very long time, Oliver and his dad reached the north shore of the bay and drove the car off the ramp. Oliver asked his dad, "Why did the ferry not get pushed out into the bay when our tires pushed forward on the ferry?" His dad replied, "Look back at the dock and see how the ferry has huge ropes wound tightly around the dock posts and the ferry's frame. Also, think about how much heavier the ferry is compared to our car."

After driving through the countryside and then entering grandpa's property, Oliver noticed a horse running through a meadow as they rounded a gentle curve around the farm. The horse seemed to be gaining on them. The horse was running straight across the field as they were driving around it. "He took a shortcut!" exclaimed Oliver. It looked like the horse had outrun them. When they finally got to the house, Grandpa came out to greet them. He asked Oliver how the trip was. Oliver thought for a moment and said, "Well, I have a lot of questions to ask my science teacher about motion!"

Describe several types of motion used in this story. _____

_____

_____

Why did Oliver's father set the brake on the ferry? _____

_____

_____

Name _____

Speed

Find a partner. Take turns being the timekeeper and being the person who moves. Your teacher has marked a specific distance for your class. In this activity, you will use the equation for speed to determine how fast you and your partner can walk, run, and crab-walk. In order to do this, you will take turns timing your partner as he or she completes the three types of motion listed in the chart. Fill in the blanks in the chart below, using the data you collect during this activity.

$$\boxed{\text{Speed} = \text{distance} \div \text{time}}$$

1. Your partner will stand at the start line. When your teacher says go, start the timer. Your partner will walk at a normal pace until he or she gets to the finish line. When he or she crosses the finish line, stop the timer and record the time.

2. Your teacher will tell you when to begin again. Repeat the above steps for running and for crab-walking. To crab-walk, place your hands and feet on the ground, with your bottom toward the ground. Then begin to move, keeping this position.

3. Switch with your partner and repeat each activity.

4. Complete the chart below, using the above equation and the collected data.

Type of Motion	Partner's time (sec)	Distance (ft)	Speed (ft/sec)	Speed (mph)
walking				
running				
crab-walking				

5. What was your partner's average time for the three different trials? _____

What was your partner's average speed in feet per second? _____

What was your partner's average speed in miles per hour? _____

6. Using your average speed in miles per hour, how far would you travel in 8 hours? _____

_____

7. A peregrine falcon flies at an average speed of 48 kph (30 mph). At that speed, how far would

it fly in 40 minutes? _____

Name _____

Speed Versus Velocity

Speed and velocity are often mistaken as the same thing. Speed is the measure of distance per unit of time. Velocity is the speed in a certain direction. Look at the map below, and use it and the equation for speed to answer the questions.

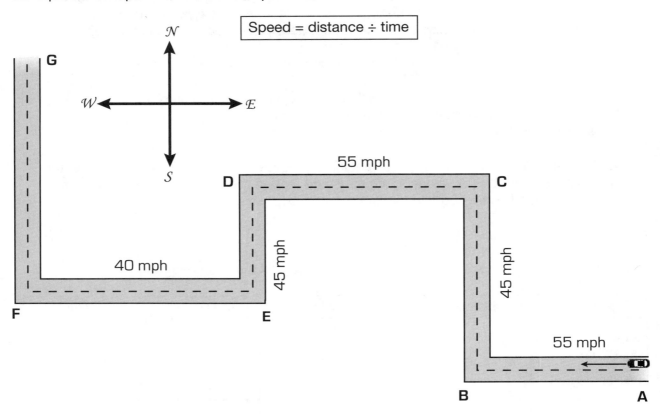

Speed = distance ÷ time

1. Give the velocity for the car traveling between the following locations:

   a. from *A* to *B*: _____

   b. from *B* to *C*: _____

   c. from *E* to *F*: _____

2. Look at the map again and compare the velocities of the car's travels.

   a. In which two segments does the car have the same velocity? _____

   b. In which two segments does the car have the same speed but not the same velocity?

   _____

3. If the distance between *F* and *G* is 83 miles and the driver travels that distance for 2.5 hours,

   what would the driver's velocity be? Show your work. _____

Name _____

Marble Mania

Work in groups of three to four students to demonstrate the Law of Conservation of Momentum. Your teacher will give you a ruler or yardstick, and 7 marbles. Follow the instructions below to complete the activity.

1. Set the ruler or yardstick on a flat surface. Tape its sides to the flat surface, making sure not to cover the center groove.

2. Place 2 marbles in the center groove, making sure they are touching each other.

3. Roll 1 marble toward the 2 marbles that are touching each other. Record your observations.

   _____

   _____

4. Now place 3 marbles in a row and repeat *Step 3*. Record your observations.

   _____

   _____

5. Repeat *Step 4* with 4 marbles in a row, and then again with 5 marbles in a row. Record your

   observations. _____

   _____

6. Repeat the experiment, except this time roll 2 marbles into the row of 4 marbles. Record your

   observations. _____

   _____

7. Predict what would happen if you rolled 3 marbles into 4 marbles.

   _____

8. Test your prediction. What were the results?

   _____

9. How does this activity demonstrate the Law of Conservation of Momentum?

   _____

   _____

   _____

   _____

   _____

Name _____

Acceleration and Momentum

Use the following equations and what you have learned about momentum and acceleration to complete the word problems. Units for acceleration are in m/s². Common units for momentum are kg×m/s and kg×kph.

> Acceleration = force ÷ mass
>
> Momentum = mass × velocity

1. A car with a mass of 1,000 kg is moving with a velocity of 6 kph north. What is its momentum?

2. A baseball with a mass of 0.2 kg is moved with a force of 10 N. What is the acceleration of the ball? _____

3. A bumper car hits another bumper car at rest. Both have the same mass. What is the resulting momentum? _____

_____

_____

4. Explain your answer to *Question 3* using the Law of Conservation of Momentum.

_____

_____

_____

5. What force is needed to accelerate a mass of 50 kg to 25 m/s²? _____

6. How are acceleration and momentum similar?

_____

_____

Name _____

Comparing Friction

Different surfaces have varying amounts of friction. For example, it is much easier to slide down a snowy hill than a dry grassy hill. Scientists measure how much force is required for two objects to move past each other when they are touching. The higher the number, the more resistance objects encounter. When two objects are made of the same material, the force required to slide past is different for static versus kinetic friction. Use the information below to draw a double line graph making a comparison of cast iron sliding over another material. Use black for static values, and red for kinetic values.

Material 1	Material 2	Static Friction	Kinetic Friction
cast iron	steel	0.40	0.23
cast iron	cast iron	1.10	0.15
cast iron	zinc	0.85	0.21
cast iron	copper	1.05	0.29

Comparisons of Frictional Force Between Cast Iron and Other Materials

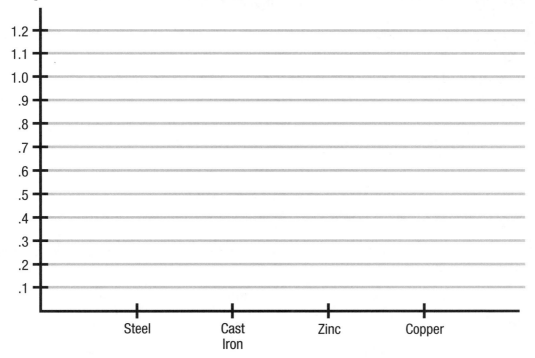

Name _____

Comparing Friction, continued

Answer the following questions using the graph you made in **Science Notebook 8.5A**.

1. Look at the black and red graph lines. Higher values mean greater resistance between objects due to friction. Based on this data, what can you conclude about the friction of moving versus stationary objects? _____

_____

_____

_____

2. Which materials have the greatest friction when moving across each other?

_____

3. Which materials have the least friction when moving across each other?

_____

4. Of the two examples below, which do you predict would have the lowest comparison value of friction?

car tires on wet road   car tires on icy road

5. Of the following two examples, which do you predict would have the highest comparison value of friction?

car stopped at a red light   car continuing through an intersection

6. Explain your answers to *Questions 4* and *5*.

_____

_____

_____

_____

7. Circle the items below that would reduce friction between two objects, and underline the ones that would increase friction.

oil	wax	glue	tape
water	honey	rubber	glass
sand	leather	grease	lotion

Name _____

Momentum and Collisions

When two objects collide and there is no other outside force in operation, each object's momentum is not lost but transferred to the other. This means that momentum is conserved. When all the momentum in one is transferred to the other, it is called an *elastic collision*. An example of this is when a moving billiard ball collides with a stationary billiard ball. Although most collisions are not completely elastic or inelastic, they can be considered more or less elastic. The more elastic a collision, the more it will rebound when objects collide. Inelastic collisions will have a smaller rebound, and may even stick together, resulting in no rebound. An example of this is when two balls of playdough are thrown together. They change shape and stick together.

Your teacher has set up three stations. At each station there is a surface of either: foam-filled poster board, magazines, or cinder block; four different types of balls; and a meterstick. Start at a station and rotate through all of them until you have completed the activity.

1. Record the type of surface.
2. Using the meterstick as your guide, hold the first ball 1 m (39 in.) above the surface and drop it. Record the distance it rebounds at the first bounce.
3. Note whether the collision is elastic or inelastic based on criteria given by your teacher.
4. Repeat *Steps 1, 2,* and *3* for the other balls.
5. Move to the next station and repeat *Steps 1–4*.

6. Before performing the test, predict whether each surface will be involved in more elastic or

inelastic collisions and explain why. _____

_____

7. Predict whether each kind of ball will be involved in more elastic or inelastic collisions and

explain why. _____

_____

Label the type of surface at each station and record your results in the table below. Use *E* to indicate an elastic collision and *I* to indicate an inelastic collision.

Ball Type	Station 1 Surface Rebound Height (cm) _____	Elasticity	Station 2 Surface Rebound Height (cm) _____	Elasticity	Station 3 Surface Rebound Height (cm) _____	Elasticity
baseball						
tennis ball						
table tennis ball						
super bouncy ball						

Name _____

Momentum and Collisions, continued

The Law of Conservation of Momentum states that momentum is not _____ or

_____. Without other _____ involved, when two objects collide,

the _____ is the _____ before and after the collision.

1. List two forces that can hinder momentum.

    a. _____   **b.** _____

2. In the activity, did any of the results surprise you? Why or why not?

_____

_____

3. Which balls were more elastic? _____

4. Which were inelastic? _____

5. Which surfaces were more elastic? _____

6. Which were inelastic? _____

7. Which ball and surface caused the most elastic collision? _____

8. Which ball and surface caused the most inelastic collision? _____

9. Choose an elastic collision and an inelastic collision and explain why each ball bounced to the

    height that it did. _____

_____

_____

10. Was momentum conserved in this activity? _____ How did you reach that conclusion?

_____

_____

11. What can you conclude from this experiment about momentum and collisions?

_____

_____

_____

_____

Name _____

Testing Friction

Follow the directions below to complete the activity. Record your predictions.

Question: How do different types of surfaces affect friction?

Predict: Which book will be the easiest to pull? _____

Which book will be the hardest to pull? _____

Try It Out:

At each station there is a book that has a string attached to it. The books are wrapped in aluminum foil, waxed paper, or a hand towel. One is left unwrapped. At each station there is also a spring scale. Hold the spring scale and attach the string from the book to it.

1. Touch the material the book is wrapped in and record your description of it.
2. Place the book down on its front cover.
3. Hold the spring scale and pull the book across the desk or table.
4. Pull gently until you are just able to drag the book across the desk. Record the force in newtons that is shown on the spring scale.
5. Turn the book on its binding, and repeat *Steps 3–4*. You may need to gently guide the book to keep it from falling over.
6. Repeat *Steps 1–5* at each station.

Covering	PHYSICAL DESCRIPTION	SIDE	SPRING SCALE MEASUREMENT
aluminum foil		front cover	
		binding	
waxed paper		front cover	
		binding	
hand towel		front cover	
		binding	
unwrapped		front cover	
		binding	

Name _____

Testing Friction, *continued*

Analyze and Conclude:

1. Were your predictions accurate? Why or Why not?

_____

_____

2. In general, was it easier to pull the book on its binding or front cover?

_____

3. How does this relate to your understanding of friction?

_____

_____

_____

4. What can you conclude about surface area and friction?

_____

_____

_____

5. What can you conclude about the different surfaces?

_____

_____

_____

6. Using the results from this experiment, explain why snow skis are coated in wax.

_____

_____

_____

7. Transporting heavy loads by railway is very efficient due to the small amount of friction between the steel wheel and the steel train track. Friction produced from steel-on-steel contact is considerably less than that of flexible rubber tires on pavement. Use your results to explain why trains would be a good form of transport as related to friction.

_____

_____

_____

Name _____

Perpetual Motion

Use complete sentences to answer the following questions:

1. What are three factors that must be overcome in order for perpetual motion to occur?

_____

_____

2. Has anyone ever succeeded in building a working perpetual motion machine? Why?

_____

_____

_____

3. What are two of the laws of thermodynamics?

_____

_____

_____

_____

4. Using the image below, describe how this machine should work if it were in perpetual motion, and how it actually works because of factors hindering perpetual motion.

_____

_____

_____

_____

_____

_____

_____

5. Why is a windmill not a perpetual motion machine?

_____

_____

_____

_____

Name _____

Thermodynamics

Thermodynamics is the study of the relationship between energy and work. The First Law of Thermodynamics states that energy is not created or destroyed; it is converted from one form into another. An example would be converting the energy of a moving object, or kinetic energy, into heat energy. The Second Law of Thermodynamics involves the idea of *entropy*. Entropy is the universal process by which all living things and all matter move from order to disorder. In other words, all things age and break down. This means that there is inefficiency when energy is converted. It also means that matter has the tendency to go to chaos and not order.

Describe what things must be taken into consideration when designing a perpetual motion machine. Consider the laws of motion and thermodynamics, and how they relate. Describe how the laws of motion would be in your favor, and how the laws of thermodynamics would be a challenge.

_____
_____
_____
_____
_____
_____
_____
_____
_____
_____
_____

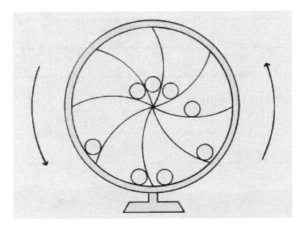

Name _____

Vocabulary Review

The bold-faced words below are paired with the wrong definition. Cross out the incorrect terms and write in the vocabulary term that belongs to the definition on the line provided.

1. average speed _____
   motion that continues endlessly

2. force _____
   the measure of an object's speed and direction

3. friction _____
   the force that resists the motion of a moving object

4. inertia _____
   the property of a moving object that causes it to tend to keep moving

5. kinetic friction _____
   the change in an object's velocity over a period of time

6. velocity _____
   the scientific study of how energy transfers from one place to another and from one form to another

7. momentum _____
   the action of friction caused by air

8. perpetual motion _____
   the push or a pull on an object

9. static friction _____
   the standard metric unit of measurement of force

10. thermodynamics _____
   the force that resists motion between two surfaces in contact with one another

11. acceleration _____
   the total distance traveled divided by the total travel time

12. air resistance _____
   the force that resists movement of an object at rest

13. newton _____
   the resistance to a change in motion

Name _____

Chapter 8 Review

Complete the exercises below.

1. Which has more momentum, a baseball of 0.2 kg traveling at 100 kph, or a 0.45 kg basketball

moving at 50 kph? _____

2. What is the difference between speed and velocity?

_____

_____

3. A girl rides her bike from her house down the street to her friend's house. When she finds out that her friend is not there, she gets on her bike and returns home. Her total distance traveled was 1.5 miles. It took her 10 minutes to complete her trip. What was her average speed?

4. How is acceleration related to force?

_____

_____

5. Calculate the acceleration of a 45 kg person pushed with a force of 10 N. _____

6. A collision can be described as *elastic* or *inelastic*. What is the difference between the two? List three factors which can affect the elasticity of a collision.

_____

_____

_____

7. On the image below, assume the man is slowly pushing the car. Draw arrows pointing to the areas where there would be friction. Label as either *static* or *kinetic friction*, and describe what would happen if friction were decreased.

_____

_____

_____

_____

8. Write Newton's three Laws of Motion.

_____

_____

_____

Name _____

Water Patterns

Obtain materials from your teacher and follow the directions below. Answer the questions as you complete the steps.

1. Your bottle contains liquid soap. Add two drops of food coloring to the soap.
2. Go to the sink and fill the bottle with water very slowly. Tip the bottle so that the water runs down the inside so as to not create suds. Fill the bottle completely.
3. Screw on the bottle cap tightly. Slowly rotate and turn the bottle upside down to mix the liquids. If suds develop, take the cap off and slowly add more water to the bottle. This will push out the suds.
4. Use the paper towels to dry off the bottle. Then wrap the tape around the cap to prevent any leakage.
5. Slowly rotate the bottle and describe what you see.

_____

_____

_____

6. Swirl the water quickly and describe what you see.

_____

_____

_____

7. Shake the bottle and describe what you see.

_____

_____

_____

8. How does what you have observed compare to ocean water you have seen and/or studied?

_____

_____

_____

Name _____

Message in a Bottle

Imagine that you are going to put a message in a bottle and throw it in the ocean to see where it will go. Study the diagram of global surface currents below. Then predict the bottle's final destination.

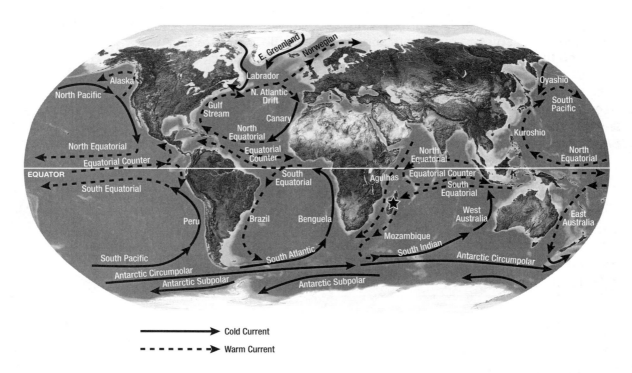

→ Cold Current

- - - → Warm Current

1. Find the star marked on the map above. This is the first location where the bottle will be put into the ocean. Use a colored pencil or marker to trace the path in which you think the bottle will travel. Mark the place you think it will land with a solid circle. Then write a description of the path and include the name(s) of the currents, continents, countries, and oceans, as well as the directions it traveled.

   _____

   _____

   _____

2. Choose a different coastal location on the globe and mark it with an *X*. Predict the path that another bottle will take from this point. Trace it with your colored pencil or marker and mark the final destination with an asterisk (*). Describe the path and include the name(s) of the currents, continents, countries, and oceans, as well as the directions it traveled.

   _____

   _____

   _____

Name _____

Current-ly

Complete and answer the following:

1. The two main types of currents are _____ and

_____ .

2. How do surface currents tend to flow? _____

3. What is the cause of longshore currents? Describe their effect on the shoreline.

_____

_____

4. Why are rip currents dangerous for swimmers? _____

_____

_____

5. If caught in a rip current, in what direction should you swim? _____

_____

6. Compare and contrast an upwelling and downwelling. Include a diagram of each event.

_____

_____

_____

7. Explain what causes deep currents and describe how they flow.

_____

_____

_____

_____

Name _____

Current Density Solution

Your task is to show how density affects ocean currents. Make sure to include temperature and salinity. Ask a question and make a prediction. Then use the materials your teacher gives you and develop a plan to test your prediction. Write out and number each step of your experiment. Use an additional sheet of paper if necessary. Then test your plan.

Question: _____

Predict: _____

_____

Try It Out: _____

_____

_____

_____

_____

_____

_____

_____

_____

_____

_____

_____

_____

_____

_____

Analyze and Conclude:

Was your prediction correct? Explain. _____

_____

Use the results of your experiment to draw a conclusion about temperature, salinity, density, and currents. _____

_____

_____

Name _____

Labeling Zones

Use the following terms in the Word Bank to label the diagram below.

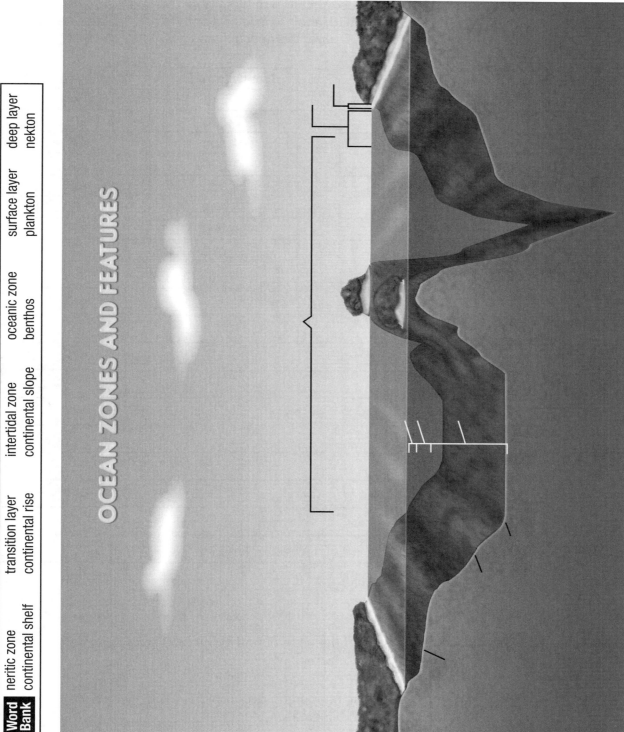

OCEAN ZONES AND FEATURES

Word Bank

neritic zone	transition layer	intertidal zone	oceanic zone	surface layer	deep layer
continental shelf	continental rise	continental slope	benthos	plankton	nekton

Name _____

Graphing Temperature Versus Depth

Using the data given, plot a line graph comparing temperature changes with depth in temperate and tropical regions of the ocean. Remember that in the ocean, the surface is considered the starting point (0 m), so the zero should be at the top of the graph. Use a different color to represent each geographical region. Be sure to label the title and both axes. Include a key to show which color represents which region. Answer the questions below.

TEMPERATE

Depth (meters)	Temperature (°C)
0	18
500	15
1,000	8
1,500	5
2,000	4
3,000	3
4,000	3
5,000	2

TROPICAL

Depth (meters)	Temperature (°C)
0	24
500	20
1,000	5
1,500	4
2,000	4
3,000	3
4,000	3
5,000	2

Title: _____

1. Between what depths is the sharpest drop in temperature seen? _____

2. What can you predict about the density of the water found at the 5,000 m depth?

   _____

3. Of the three major climate regions—polar, temperate, and tropical—which would you predict would change the least in temperature with increasing depth?

   _____

Name _____

Identifying Marine Life

Write the marine organism groups—one in each small box below—and briefly describe each group on the lines. Then observe the pictures. Identify which group the organism belongs to and explain why it belongs in that group. Write your answers on the lines below each picture.

Marine Organism Groups

Name _____

Underwater Uniqueness

Read the paragraphs below and answer the following questions:

In order to survive in their environment, some animals in the deep zone have a definite advantage. They can produce light, or bioluminescence, by using chemicals within their own bodies. Bioluminescence is mainly a marine phenomenon, but is not limited to ocean-dwelling animals. A few land animals, such as the firefly, have this ability also. Some marine organisms give off the light continually. Others flash it on and off. Bioluminescence is used by certain organisms to communicate and attract mates. Some use it as a warning sign to stay away or as a form of camouflage. Still others may use it for navigation or to attract prey. An excellent example of this is the anglerfish. This peculiar-looking fish fits perfectly into its undersea niche by luring other deep-sea fish to itself with the help of a glowing lure. When its prey gets close enough, the anglerfish attacks with its lightning-fast jaws and enjoys a tasty meal!

For many animals, the landscape of their habitat protects them and helps them to survive. Examining a coral reef provides evidence of organisms, both plant and animal, that are found in symbiotic relationships. A symbiotic relationship is one that benefits one or both organisms. For example, some types of coral require a special type of algae to provide much of their energy. The algae use the reef corals as dwelling places. Another example is clownfish, which hide among the tentacles of sea anemones when danger is near. The sea anemones rely on the clownfish to bring back food to eat. This mutually beneficial relationship allows both organisms to survive.

1. Why is bioluminescence unique?

_____

_____

_____

2. Explain why bioluminescence is a characteristic that helps deep-sea creatures survive.

_____

_____

3. How are symbiotic relationships helpful?

_____

_____

4. Pick one of the examples of symbiosis discussed above or choose one of your own examples. Describe how one or both of the organisms benefits the other.

_____

_____

Name _____

Measuring Salinity

The amount of salt found in different areas of ocean water affects how the currents flow. In some areas of the ocean, very dense, salty water, also known as *brine*, forms when extremely large amounts of salt are dissolved in a small quantity of water. Hydrometers are instruments that measure the density of water. You are going to construct a hydrometer and then use it to estimate the salinity level of different saltwater solutions.

Procedure:
1. Obtain materials from your teacher. You will also need to obtain five different solutions from your teacher during the experiment.

2. Lay the straw next to the millimeter side of the ruler. With the marker, carefully make lines on the side of the straw every 5 mm, from one end to the other. Do this on all three straws. Label one straw with the number *1*, the second straw with a *2*, and the third straw with a *3*.

3. On your piece of waxed paper, divide the clay into three pieces. Roll each piece into a small ball. Plug one end of each of the three straws with the clay, carefully molding it tightly around the straw. Use the smallest amount of clay possible that will completely cover the bottom of the straw. If you use too much, your hydrometers will be too heavy and will not work. You should now have three hydrometers.

4. Pour 200 mL of room temperature tap water into your beaker. Test the hydrometers by placing them into your beaker of water with plugged end down. If the straws are not able to stay upright in the water, drop some small metal beads or BBs into the straws, one at a time, until they stay upright. It is very important that your straw float in the beaker without touching the bottom or sides of the container.

5. Begin with *Straw 1* in the tap water only. Count the number of marks that are submerged. If the water level falls between lines, estimate the measurement, such as 4.5 lines. Remember that each line represents 5 mm. Multiply the number of lines by 5 to calculate the total number of millimeters. For example, if the water level is 4.5 lines, multiply by 5. The measurement of the water level would be 22.5 mm. Round to the nearest whole number, which for this example would be 23 mm. Record this distance in the chart on *Data Table 1*. Do this for *Straw 2* and *Straw 3*. Record the measurements on the data table.

6. Calculate the average of the three distances and record this on *Data Table 1*. Round to the nearest whole number.

7. Pour out the tap water in your beaker and replace it with 200 mL of 100% solution. Repeat *Steps 5* and *6* for this solution. Record the measurements on *Data Table 1*.

8. Pour out the 100% solution and replace it with 200 mL of 50% solution. Repeat *Steps 5* and *6* for this solution. Record the measurements on *Data Table 1*.

9. Pour out the 50% solution and replace with 200 mL of 25% solution. Repeat *Steps 5* and *6* for this solution. Record the measurements on *Data Table 1*.

Name _____

Measuring Salinity, continued

DATA TABLE 1

Floating Heights (mm)

Salt concentration	Straw 1	Straw 2	Straw 3	Average
Tap water (0%)				
Saturated (100%)				
Half saturated (50%)				
One-fourth saturated (25%)				

10. Find the graph on the next page. Plot the average heights for each salt concentration level from *Data Table 1*. Construct a line graph to show how the depth changes with different levels of salt content.

11. Empty your beaker and replace it with 200 mL of the mystery test solution.

12. Place *Straw 1* in the mystery test solution. Observe and calculate the water level height. Record this measurement in *Data Table 2*.

13. Repeat *Step 12* for *Straws 2* and *3*.

14. Find the average of the three measurements and record it. Plot the average measurement on the graph. Predict the salt concentration of the mystery test solution by comparing it to the other solutions. Record your prediction in the Salt Concentration box in *Data Table 2*.

DATA TABLE 2

Floating Heights (mm)

Straw 1	Straw 2	Straw 3	Average	Salt Concentration

Put all materials and equipment away as directed by your teacher. Be sure your work area is clean before proceeding. Use the data from the charts to answer the questions on the next two pages.

Name _____

Analyzing Measurements

GRAPH

Title: _____

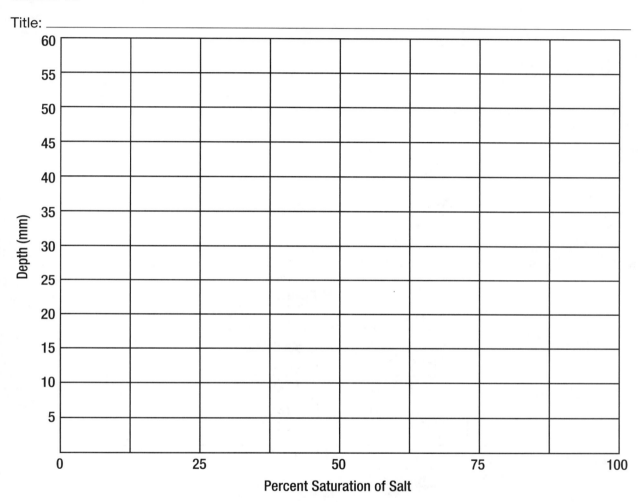

Depth (mm) (y-axis): 60, 55, 50, 45, 40, 35, 30, 25, 20, 15, 10, 5

Percent Saturation of Salt (x-axis): 0, 25, 50, 75, 100

1. How does the concentration of salt in a solution affect the height of a floating hydrometer?

_____

_____

2. Based on the data you gathered, where would it be easier for you to float—in a lake or an ocean?

_____

3. Explain why.

_____

_____

Name _____

Analyzing Measurements, continued

4. Based on your graphed data, estimate what the floating height of a straw in millimeters would be in a solution that is:

75% saturated _____

90% saturated _____

5. Based on your graphed data, what would the saturation percent of a solution be if the hydrometer floated at a height of:

37 mm _____

33 mm _____

25 mm _____

Use the density table below to answer the following questions about liquids and their densities:

DENSITY TABLE

Substance	Density (g/cm^3)
gasoline	0.70
ice (at 0° C)	0.92
water (at 4° C)	1.000
seawater	1.026
milk	1.03
mercury	13.6

6. Which substances are less dense than water? _____

7. Which substances listed would float in water? _____

8. What would happen if mercury were put in water? Why?

_____

_____

9. Do the results of your experiment verify the information on the density table concerning the differences between freshwater and seawater? Explain why or why not.

_____

_____

Name _____

Graphing Salinity Versus Depth

Imagine that you are a scientist studying a tropical area of the ocean that has not yet been measured or discovered. You are specifically collecting data on salinity levels. Normal ocean water ranges between 3.0–3.8% salt content. Graph the data on a line graph. Include a title and label both axes. Remember to start with the zero at the top of the graph. Analyze the graph and answer the questions.

Depth (m)	Salt Content (%)
0	3.2
200	3.3
400	3.3
600	3.4
800	3.5
1,000	3.5
1,200	3.3
1,400	4.4

Title: _____

1. At what depth did the salinity begin to decrease significantly? _____

2. At what depth is the salinity the lowest? _____

3. At what depth is the salinity the highest? _____

4. Between what two depths is the largest change recorded? _____

5. What oceanographic feature might be the cause of this? _____

Name _____

Environments

Research the organism that your teacher assigned to your group. Find the type of habitat for which it is best suited and record it. Then list any special characteristics that allow it to survive in its environment. The first one has been done as an example. Take notes on the other organisms as each group reports on their findings.

ORGANISM	ENVIRONMENT	SPECIAL CHARACTERISTICS
Bull Shark Example:	The bull shark must live in water. It can survive in both saltwater and freshwater.	The bull shark survives by controlling its own internal salt and water levels so that it can live in different environments. Its preferred environment is saltwater, however.
Green Sea Turtle		
Brown Algae (Kelp)		
Sea Lions		
Coral		
Tubeworms		

Name _____

Vocabulary Review

Use the clues below each set of blanks to fill in the corresponding vocabulary word. When you are finished, unscramble the circled letters to find the hidden word.

1. _____ ◯ _____ _____ _____ _____

the continuous flow of ocean water in a certain direction

2. _____ _____ _____ _____ _____ ◯ _____ _____ _____

the zone located between high- and low-tide lines

3. ◯ _____ _____ _____ _____ _____

the organisms that inhabit the ocean floor

4. ◯ _____ _____ _____ _____ _____ _____ _____ _____ _____ _____ ◯ _____

the movement of water caused by waves that strike the shore at an angle

5. _____ _____ _____ _____ ◯ _____ _____ _____ ◯ _____ _____ _____

a gently sloping area of the ocean floor that extends outward from the shoreline

6. _____ _____ _____ _____ _____ ◯ _____ _____

an area of water on the ocean floor that has high salinity

7. _____ ◯ _____ _____ ◯ _____

free-swimming animals

8. _____ _____ ◯ _____ ◯ _____

the zone that extends from the low-tide line out to the edge of the continental shelf

9. _____ _____ _____ _____ ◯ _____

tiny algae and animals that float on the water's surface

10. _____ ◯ _____ _____ _____ _____

the zone that extends from the edge of the continental shelf and covers the open ocean

---

11. The hidden word is

_____ _____ _____ _____ **M** _____ _____ _____ _____ _____ _____ _____

12. What is the hidden word's definition?

_____

_____

Name _____

Chapter 9 Review

Complete the following exercises:

1. Draw and label a diagram in the box that represents the temperature layers of the ocean.

2. What are the three basic ways that ocean water moves?

   a. _____

   b. _____

   c. _____

3. What is an upwelling? Why is it beneficial? _____

   _____

   _____

4. Describe how waves begin and move through the water to the shore.

   _____

   _____

   _____

   _____

5. What causes surface currents and what pattern do they follow? _____

   _____

6. How do rip currents form? _____

   _____

7. Why do deep currents form? _____

   _____

8. Compare and contrast El Niño and La Niña.

   _____

   _____

   _____

9. What are extremophiles and how are they unique? _____

   _____

   _____

Name _____

Making a Rock

Use the materials you are given to make a rock. Follow the directions to complete the activity. Answer the questions below.

1. Pour 50 mL ($\frac{1}{4}$ cup) nontoxic plaster and 100 mL ($\frac{1}{2}$ cup) sand into a beaker and thoroughly mix them together. Carefully sprinkle the sand and plaster mixture over the water in the STYROFOAM® cup. Gently stir it with a plastic spoon. Wait about 2–3 minutes to let the mixture settle to the bottom. Do not disturb the cup after the mixture has settled.

What does this mixture represent? _____

2. Pour 50 mL ($\frac{1}{4}$ cup) plaster and 100 mL ($\frac{1}{2}$ cup) dark-colored gravel into the beaker and mix them together. When the sand and plaster mixture has settled, gently sprinkle the gravel and plaster mixture onto the water and allow time for it to settle. Remember not to disturb the cup.

What does this mixture represent? _____

3. Pour 50 mL ($\frac{1}{4}$ cup) plaster and 100 mL ($\frac{1}{2}$ cup) crushed seashells into the beaker and mix together. When the second layer has settled, gently sprinkle the shell and plaster mixture into the water and let it settle. If any of the mixture does not sink below the water, very carefully spread it out with the spoon until the mixture is submerged. Do not move the cup or try to pour out any water at this time. Let the rock mixture harden for 20–30 minutes.

Predict what kind of rock this is. _____

4. After the mixture has hardened, carefully pour out any water that may be remaining on top. Use the edge of one scissor blade to score the side of the cup. Peel away the STYROFOAM® to expose the rock.

How many layers can you see? _____

5. How is this process similar to the way in which a sedimentary rock is made naturally?

_____

_____

_____

6. How is this process different?

_____

_____

_____

Name _____

Classifying Rocks

The 20 rocks that your teacher has given you need to be classified. Work with your group to decide how to divide the rocks into groups. Follow the directions below to complete the exercises. Answer the questions.

1. Place all the rocks in a pile toward one end of the butcher paper. Draw a circle around the pile.

2. Divide the light-colored rocks into one group and the dark-colored rocks into another group. Draw a circle around each group.

3. Continue to sort the light-colored rocks into smaller groups. What criteria did you use to

   classify them? _____

   How many groups did you divide the light-colored rocks into? _____

4. Repeat *Steps 2–3* for the dark-colored rocks. What criteria did you use?

   _____

   How many groups did you make? _____

5. Take the rocks off the paper and turn the paper over. Place the rocks into one pile and draw a circle around it.

6. Divide the rocks into two piles. Choose different criteria than previously used.

   What are your two groups? _____

7. Further sort the rocks into smaller piles. How many total groups did you get? _____

   Describe the characteristics of each group. _____

   _____

8. Partner with another group of students. Take turns trying to determine what system of classification the other group used. Were you able to guess their system?

   _____

9. Look over the list your teacher wrote on the board. Did anyone classify the rocks using the same criteria? Were the rocks grouped in the same way? Explain.

   _____

   _____

10. Make a conclusion about classifying rocks.

    _____

    _____

Name _____

Agents of Breakdown and Erosion

Read the following journal entries about someone's travels. Circle any causes of weathering or erosion, and underline the verbs that describe how the forces caused the weathering or erosion.

Day 1: I visited the Carlsbad Caverns National Park in New Mexico. The tour guide informed me that through the years, slightly acidic rainwater had seeped into the ground and dissolved portions of the limestone. The rainwater also dissolved hydrogen sulfide gas coming up from the ground below. This created a sulfuric acid solution that dissolved the limestone further, producing these majestic caverns. I got a little uncomfortable in those deep, dark passageways, but enjoyed the awesome underground spaces and strange formations.

Day 5: I visited some of the great canyons in the southwestern United States, including the Grand Canyon. Park rangers told me its formation took 5 to 6 million years, but I found an interesting book in the visitor center that gave scientific evidence for a much shorter time of formation. In either case, it is evident that water has carved out huge amounts of rock, as the canyon averages 1,219 m (4,000 ft) in depth for over 435 km (270 mi)!

Day 17: I enjoy driving in wide-open spaces and mountainous regions. One time I had to re-route my trip almost a hundred miles when I encountered a road closure. Apparently a recent earthquake had shaken loose a bunch of rock from a cliff along the highway. Boulders and chunks of rock completely covered both lanes. I was disappointed at first because it made for a long day of driving. Later though, I realized how grateful I was that I had not been on the road when the rocks came crashing down.

Day 21: I completed a long hike through an amazing slot canyon here in Utah. The red and orange cliffs with their wavy stripes mesmerized me. The stripes looked smoother than the ones on the wind-blasted columns I saw earlier on the same trip. I also read a pamphlet that described how flash floods from thunderstorms could quickly fill these slots with rushing water. Apparently that is how these particular slot canyons were carved to such dramatic depths.

Day 26: I had the opportunity to hike in Glacier National Park, Montana. I found it amazing how ice, under the force of gravity, carved out such beautiful lakes and hanging valleys, leaving behind large hills called *moraines*. When I got up close, I also noticed that ice, rock, and gravity also work together to scour fine lines, referred to as "striae," in the bedrock. These striae look like stripes of color or texture.

Day 30: Soon I plan to visit some of the beautiful sand dunes on Oregon's coast. I am told that steady winds actually cause these sand dunes to migrate in one direction, and move like slow-motion waves. I believe God is constantly reshaping his Creation. I love to see and learn about all His marvelous works.

Name _____

Rock On

Read and complete the following exercises.

1. What are the two types of Earth's crust? _____

2. Describe minerals. List four details. _____

   _____

3. Explain how the formation of an igneous rock affects its grain texture.

   _____

   _____

4. What are the contributing factors that produce metamorphic rocks? _____

   _____

5. Describe how a rock becomes foliated. _____

   _____

6. What type of change occurs when metamorphic rock is transformed by heat? _____

   By pressure? _____ By heat and pressure? _____

7. In what two ways can sedimentary rock form? _____

   _____

   _____

8. Flowing water carries a _____, which is dissolved or _____ matter.

9. Describe two ways in which water erodes and carries away rock. _____

   _____

   _____

10. Explain how wind causes rock to break down.

    _____

    _____

    _____

11. What are the three ways in which gravity causes erosion?

    _____

    _____

Name _____

Observing Sand

Sand is a naturally occurring, granular material made of rock and mineral particles. An individual particle of sand is called a *sand grain*, of which the diameter can range from 0.0625 mm to 2 mm (less than $\frac{1}{10}$ in.). Silt consists of particles that are smaller than sand. Particles that range from 2 to 64 mm ($\frac{1}{10}$ to $2\frac{1}{2}$ in.) are considered gravel. The composition of sand is highly variable although it usually consists of silica, commonly in the form of quartz. The color of sand varies, depending on the local rock sources. Rock is eroded and transported by water and wind and is deposited in beaches, dunes, sand spits, and sand bars.

Observe the different types of sand your teacher gives you. Classify the samples according to particle size. Then follow the directions for a sand deposition experiment and answer the questions.

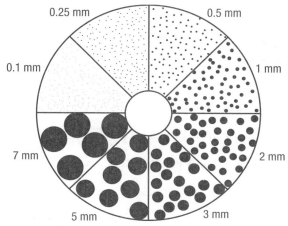

1. Get three different samples of sand from your teacher. Carefully pour one sample onto a piece of paper. Find the smallest and the largest particle. Find the size of the particles by placing them onto the grain size chart. Move the grains around the circle until you match the correct size.
Sample 1:

What is the size of the smallest particle? _____ the largest? _____

Describe the color(s) of the sand. _____

2. Put the sand back in its container. Repeat *Step 1* for the other two samples. Be careful not to get the different samples mixed together. Record the sizes for each sample in the blanks below.
Sample 2:

What is the size of the smallest particle? _____ the largest? _____

Describe the color(s) of the sand. _____

Sample 3:

What is the size of the smallest particle? _____ the largest? _____

Describe the color(s) of the sand. _____

Name _____

Observing Sand, continued

3. From the sample of sand that has many different grain sizes in it, pick out three more grains. Find the shapes of all five grains using the chart below. Write the number of the shape that most closely resembles each grain.

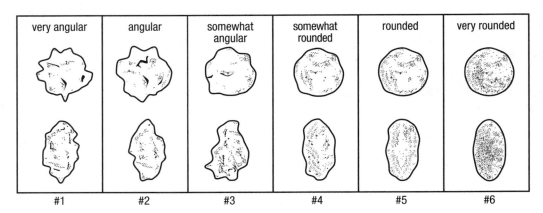

very angular	angular	somewhat angular	somewhat rounded	rounded	very rounded
#1	#2	#3	#4	#5	#6

Sample 1: _____ **Sample 2:** _____ **Sample 3:** _____

Sample 4: _____ **Sample 5:** _____

4. Does all sand have to be the same size and shape? _____ Why do you think the

grains vary?_____

5. Thoroughly mix all three samples of sand in a jar. Pour enough water into the jar so that the sand circulates easily. Screw the lid on tightly and swirl the jar around so that the sand is definitely suspended in the water. Set the jar down and allow the particles to settle.

6. Describe what happened to the particles after they settled.

_____

_____

7. Draw what you see in the jar.

8. Why do you think the sand settled in this way? Use the term *deposition* in your answer.

_____

_____

_____

Name _____

Making Fossils

Scientists search for and study fossils to learn about the past. Fossils can form in a number of ways, which causes different types of fossils. Follow the directions below to model how three different kinds of fossils are made.

1. Pour enough pre-mixed nontoxic plaster into a paper bowl so that it is about $1\frac{1}{4}$ cm ($\frac{1}{2}$ in.) thick.

2. Choose three or four small specimens, such as twigs, bone, shells or plastic animals and plants. Use a paintbrush to coat two of the objects with mineral oil.

3. Gently press the specimens into the nontoxic plaster. Allow the plaster to dry for about 10–15 minutes, or until hard.

4. While you are waiting for the plaster to dry, get a plum-sized ball of artist's modeling clay. Slightly flatten the clay, leaving it at least $1\frac{1}{4}$ cm ($\frac{1}{2}$ in.) thick.

5. Coat the remaining specimens with mineral oil and press them into the clay. Make sure to press them down so the indentations are deep enough to fill with plaster. Remove the specimens and observe the details of your fossils.

6. Do your simulated fossils in *Step 5* have the same details and patterns as your specimens? Explain the reason(s) for this.

_____

_____

7. What type of fossil did you make in *Step 5*? _____

8. Use the paintbrush to coat the indentations in the clay (made in *Step 5*) with mineral oil. Carefully fill each indentation to the top with plaster. Do not overfill. Allow the plaster to dry for about 10 minutes.

9. Check the first set of fossils from *Step 3* to see if they have sufficiently hardened. If the plaster is firm to the touch and does not move, take out or peel off the specimens. Observe the details.

10. Are the details and patterns similar to the specimens? Explain.

_____

_____

11. What type of fossil did you make in *Step 3*? _____

Name _____

Making Fossils, continued

12. Check the third set of fossils (made in *Step 8*) to see if they have hardened. When they are firm and solid to the touch, remove the dried plaster pieces from the clay. Observe the fossils.

13. Are the details and patterns similar to the specimens? Explain.

_____

_____

14. What type of fossil are these? _____

15. Compare and contrast these fossils to those made by pushing the specimens into the clay.

_____

_____

_____

16. Use the materials you have available to devise a plan to make a simulated trace fossil. Include what kind of specimen you will use. Write down your plan.

_____

_____

_____

17. Try out your plan. Did it work? _____

18. Describe what your fossil looks like and explain why it would be considered a trace fossil.

_____

_____

_____

19. Amber is the hardened resin, or sap, from plants. If an insect gets caught in the resin, dies,

and becomes covered with more resin, it becomes a _____.

20. Sometimes dissolved minerals fill in spaces of an organism. A process known as

_____ occurs if the minerals crystallize and replace all or

part of the organism.

21. Give an example of a fossil that is formed by the process mentioned in *Question 20*.

_____

Name _____

Creation Time Line

Collect the materials from your teacher. Make a time line to represent the days of Creation, gluing the pictures in the appropriate places to illustrate what God created each day. If you cannot find a suitable picture, draw one instead. Label the days and list what God created on each day according to the information in your textbook.

Name _____

Evolution Time Line

Make a time line to represent the origin of man according to evolutionists. Glue or draw pictures in the appropriate places to illustrate your time line. Label periods of time to show the span between ancestors and descendents.

Name _____

Stepping to the Past

Trace fossils are often animal footprints. Imagine that you just found over 200 dinosaur footprints in a field where you were playing. The paleontologists that came to analyze them said the footprints were from a combination of herbivorous and carnivorous dinosaurs. This has increased your curiosity and you want to know more.

1. The statements below are out of order. Read each one and then arrange them in the order that best describes how dinosaur footprints might become fossils. Write out the statements on the lines.

   • Dinosaurs stomp through mud and leave their footprints behind.
   • A long time passes.
   • Sediment and water cover the footprints.
   • Fossilized footprints form in sedimentary rock.
   • Dinosaurs step in some soft mud and sink.
   • Many dinosaurs of various types are present.
   • Layers of sediment build up over the footprints.
   • Dinosaurs seek food where both vegetation and other animals are plentiful.
   • The sediment becomes cemented together and hardens.

   a. _____

   b. _____

   c. _____

   d. _____

   e. _____

   f. _____

   g. _____

   h. _____

   i. _____

2. Observe the footprints from a trace fossil below. Make a conclusion about what the animal was doing.

   _____

   _____

Name _____

Mapping a Mountain

Geological features and terrain can be represented on two-dimensional maps by using contour lines. Each contour line joins points of equal elevation. Follow the directions below to make a model of a mountain and then draw contour lines to map it.

1. Use the clay to construct a cone-shaped mountain at least 10 cm (4 in.) high, but no taller than the container. Make one side of the mountain a good deal steeper than the other. Also make the base of the mountain so it will fit inside the container. Draw a profile of your mountain.

2. Hold a ruler against the side of the container. Mark the outside of the container with short lines at 2.5 cm (1 in.) intervals. Place the clay mountain into the container. Add water to the first mark.

3. Use a black, permanent, fine-point marker to draw a line around the mountain at the level of the water. Add water to the second marked interval on the outside of the container. Draw a line around the mountain.

4. Repeat this procedure for each interval until the top of the mountain is reached. Cover the container with a piece of glass. Place a piece of tracing paper on the glass. Looking straight down onto the mountain, trace each line starting with the base, where it touches the bottom of the container.

5. What are these lines called? _____

6. Use the following key to mark the elevation of each contour line: 2.5 cm (1 in.) = 7 m (23 ft). Start at the base and mark it with a zero (0). Then mark each line, increasing in increments according to the key.

   How tall is your mountain? _____

7. Explain how the contour lines show which side of the mountain is steeper.

   _____

8. What do contour lines that are spread farther apart represent?

   _____

9. Compare the profile that you drew to the contour lines of your mountain. How do the lines help you understand what the mountain's shape is?

   _____

   _____

Name _____

Interpreting the Geologic Column

Although the standard geologic column's eons, eras, periods, and epochs are today described along with the geologic time scale, they can still be learned and described as *systems*. Knowing the major systems allows all geologists, no matter what interpretation, to agree as to which type of rock they are referring. Read the summary of the systems on **BLM 10.8B Standard Geologic Column Analysis** and use your textbook to answer the following questions.

1. Devise a silly acrostic sentence or sentences to help you remember the first letter of each major system, from lowest to highest:

   p _____ c _____ o _____ s _____

   d _____ c _____ p _____ t _____

   j _____ c _____ t _____ q _____

2. The oldest and longest span of time is the _____.

3. *Precambrian*: What organisms are found in this system? _____

   What type of rock dominates? _____

4. *Cambrian*: Where did the creatures of the Cambrian live? _____

   What type of organisms were in great variety? _____

   What organisms were buried earliest in the Flood? _____

5. *Ordovician*: What body feature is assumed to have evolved during this period? _____

6. *Silurian*: What land organisms show up in this system? _____

7. *Devonian*: What water and land-dwelling creatures appear? _____

8. *Carboniferous*: This system is characterized by _____ and

   _____ deposits.

Name _____

Interpreting the Geologic Column, continued

9. *Permian*: This system shows an increased number of _____ and

_____.

10. *Triassic*: This system contains abundant plants known as _____.

What type of geologic activity is evident in this system? _____

11. *Jurassic*: What extinct creatures are dominant in this system and the system above it?

_____

What are pterosaurs considered to be? _____

12. *Cretaceous*: What type of plants are found buried in this system? _____

This system is often associated with a mass _____ event.

13. *Tertiary*: What age is this the beginning of? _____

What types of creatures are found in this system more than in previous ones?

_____ and _____

14. *Quaternary*: What geologic activity characterizes this system? _____

_____

15. Based on this summary, what systems are interpreted by many catastrophists as representing

the events of the Great Flood? _____ through _____

16. In which system do humans appear? _____

17. In what two ways has the Cambrian explosion been interpreted?

_____

_____

_____

Name _____

Vocabulary Review

Complete the exercises below.

1. Listed below are sets of words. Describe the relationship between the two words by writing one or two sentences for each set.

deposition – load

_____

_____

grain – foliated

_____

_____

geologic column – geologic time scale

_____

_____

2. Identify the type of fossil and write the vocabulary term on the lines below each picture.

_____ _____ _____ _____

3. Review all 12 vocabulary words. Below, list eight of the terms that depict factual or direct evidence of Earth's history. Use the four remaining terms to answer *Questions 4* and *5*.

_____   _____   _____

_____   _____   _____

_____   _____

4. A(n) _____ believes in the biblical record of the beginning of the universe. A(n)

_____ believes that the universe developed over a great amount of time from existing matter and energy.

5. The _____ is a chart or diagram depicting the history of the

earth in chronological order, and the _____ is a column of Earth's strata that provides insight into the earth's eras and periods.

Name _____

Chapter 10 Review

Answer the following:

1. Name the three basic types of rock and the main factor(s) that make for variations within that rock type.

    a. _____

    b. _____

    c. _____

2. How do breakdown and erosion affect load and deposition?

_____

_____

3. Explain the basic differences between uniformitarianism and catastrophism.

_____

_____

4. What two major events are important to catastrophists who believe in the biblical account of Earth's history?

_____

5. Summarize how neo-catastrophism is influenced by uniformitarianism and catastrophism.

_____

_____

6. List the four fossil types in order, beginning with the one that is the most complete to the type that shows the least evidence of the original organism.

_____

7. List the twelve major rock systems, or periods, from lowest to highest.

_____

_____

_____

8. What is significant about the Cambrian and Cretaceous periods?

_____

_____

Name _____

Balancing Act

When we walk, there are many different forces at work. These forces interact in order to allow our bodies to move and stay in balance. As you perform the activities below, notice how your body responds to the different forces pulling on or supporting it. Make observations about your personal experience when you walk across the beam and record them on **BLM 11.1A Balance Observations**. Answer the questions at the bottom of this notebook page.

1. Divide into groups of three. Assign each person a role. Each person will have a chance to perform all three roles.

   walker— walks on the raised surface and between the other two individuals
   left spotter— walks on the left-hand side of the walker and provides support or resistance for the walker, as directed
   right spotter— walks on the right-hand side of the walker and provides support or resistance for the walker, as directed

2. The walker will stand on one end of the beam and walk slowly to the other end. The left spotter should be about 0.5 m ($\frac{1}{2}$ yd) to the left of the walker and the right spotter should be about 0.5 m ($\frac{1}{2}$ yd) to the right of the walker.

The following directions are given to the walker:

3. Stand on the beam and walk to the other end with your arms at your side. Notice how difficult it is and what you have to do to stay balanced. Record your observations.

4. Walk on the beam with your arms straight out to the sides. Describe the level of difficulty and how you remained balanced. Record your observations.

5. Hold the broomstick or mop handle in the center so both sides are of equal distance and then walk on the beam. Compare this to the other two trials and record your observations.

6. Walk with the stick held in the center. This time, have the left spotter gently pull down on the left end of the stick. Try to remain balanced. Record what you did to stay balanced.

7. Have the left spotter let go of the stick. Walk holding the stick in the center while the right spotter gently pulls down on the right end of the stick. Record how you remained balanced.

8. What force was always present that pulled you toward the ground?

   _____

9. What parts of your body applied forces that allowed you to stand, walk, and remain balanced?

   _____

10. When the resistance force was added to either the left or right side, what did you have to do to remain balanced? _____

    _____

Name _____

Plate Movement

There are over 20 plates, or sections, of the earth's crust that ride on the upper mantle of the earth. Since the mantle is not a solid, the plates move on top of it and interact with each other. These places of interaction are called *plate boundaries*. Read the descriptions and complete the exercises below.

1. _____ boundaries occur when two plates collide. At collision, the density of the plates determines what will happen to the plates. If the colliding plates have basically the same density, the crust of both plates is forced upward and mountains form. On the other hand, if oceanic crust collides with a plate made of continental crust, the more dense oceanic crust is forced underneath the less dense continental crust in a process known as *subduction*. Old crust is destroyed during subduction.

2. _____ boundaries are the result of two plates moving apart, or diverging. Often magma oozes up and fills in the gap, forming new crust. This type of boundary commonly occurs along the mid-ocean ridge where seafloor spreading exists.

3. _____ boundaries can cause small cracks or large faults in the crust. In this type of boundary, the plates slide horizontally past one another as they move in opposite directions. In this type of boundary, crust is not being created or destroyed. Instead, earthquake activity often takes place.

4. Label the following diagrams with the correct plate boundary type.

_____   _____   _____

5. What geologic feature other than mountains is caused by convergent boundaries?

_____

6. What two geologic features occur along divergent boundaries?

_____

Name _____

Types of Volcanoes

1. Label the following volcano diagram using the words in the Word Bank. Write your answers on the lines provided.

Word Bank	magma chamber lava	pipe ash and dust	vent crater

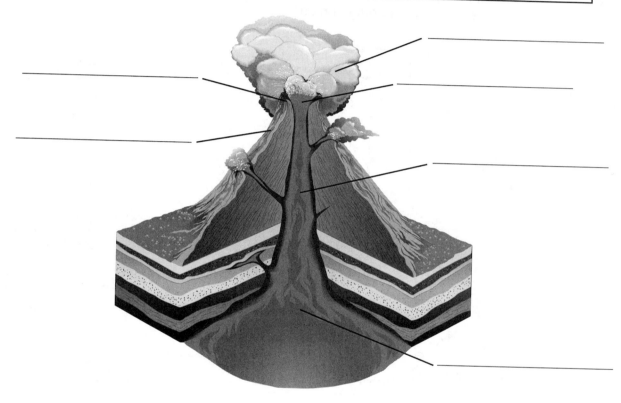

2. Compare the three basic types of volcanoes in the chart below. Write at least two things that make each volcano type unique.

SHIELD VOLCANO	CINDER CONE	COMPOSITE VOLCANO

Name _____

Active Volcanoes

Using an encyclopedia or reliable Internet resource, identify three active volcanoes that are different in type. One must be located along the Pacific Ring of Fire.

1. Mark the locations of these volcanoes by drawing an *X* to show their position on the map.

2. Record their name; their type—*shield*, *composite*, or *cinder*; and their location on the table below. Include an interesting fact about each one.

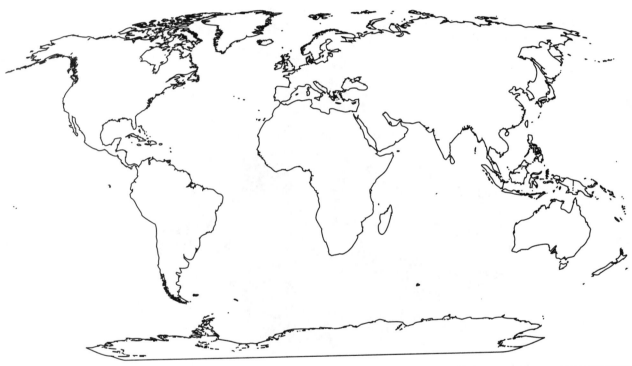

Volcano Name	Type	Location	Interesting Fact

Name _____

Tornado Frequency

Observe the map below of the United States. The number inside each state represents the average number of tornadoes per year per 26,000 km² (10,000 mi²) that occurred in that state between the years 1953 and 2004. Determine the top seven states that have experienced the greatest number of tornadoes. If needed, use an additional map to identify the names of each state. Write the state names, their abbreviations, and the average number of tornadoes in the table below.

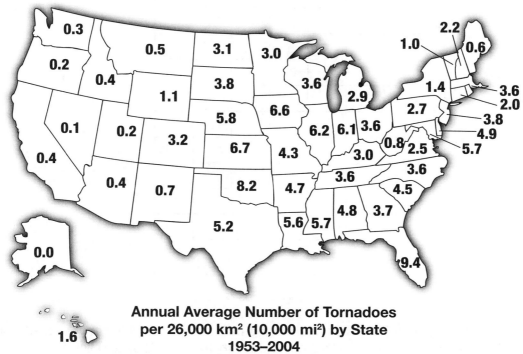

**Annual Average Number of Tornadoes
per 26,000 km² (10,000 mi²) by State
1953–2004**

Fill in the table with the correct information from the states that experience the most tornadoes, in order from greatest number to least.

State name	State abbreviation	Annual average number of tornadoes
1.		
2.		
3.		
4.		
5.		
6.		
7.		

Name _____

Graphing and Analyzing

Construct a bar graph using the data from the chart on **Science Notebook 11.4A Tornado Frequency**. Label each axis with the appropriate terms and give the graph a title. Use the state abbreviations. Plot the average number of tornadoes for each of the seven states and fill in the bars.

Title: _____

_____

1. Which state has the greatest number of tornadoes? _____

2. Compare the seven states from your chart to the map of Tornado Alley in your textbook. Find the states listed on your chart that coincide with Tornado Alley and list them.

_____

_____

3. Speculate why there are not more states from the chart found in Tornado Alley.

_____

_____

4. List the five states that experienced the fewest number of tornadoes.

_____   _____

_____   _____

_____

5. Why do you think these states had the fewest tornadoes?

_____

_____

Name _____

Hurricane Heyday

Read and fill in the blanks with the correct information.

1. Find the four storms on the map below. Determine their location and identify whether the storm should be named a *typhoon*, *hurricane*, or *willy-willy*. Write the name of the storm on the blank next to the corresponding letter.

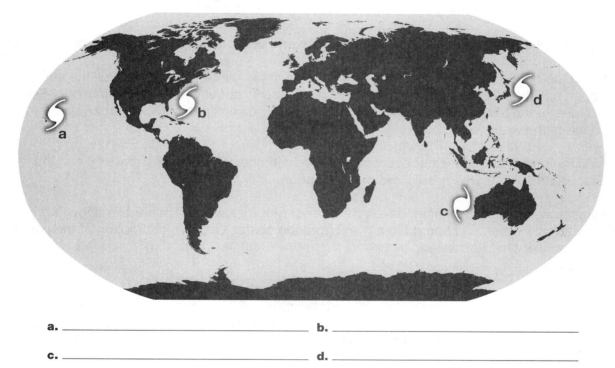

a. _____ b. _____

c. _____ d. _____

2. On the lines below, list 10 characteristics of tropical cyclones.

a. _____

b. _____

c. _____

d. _____

e. _____

f. _____

g. _____

h. _____

i. _____

j. _____

Name _____

Eye of the Hurricane

An area of low pressure forms in the center of a tropical cyclone, or hurricane, known as the *eye*. The eye is surrounded by thick cumulonimbus clouds and very intense thunderstorms, called the *eye wall*. As the outer arms, or rain bands, of the storm rotate, the eye remains calm. Read the directions below and perform the experiment to demonstrate this concept.

1. Find a partner. Obtain the materials for this activity from your teacher.

2. Fill the bowl three-fourths full with water. Tie one end of the string to the paper clip.

3. Thread the other end of the string through the hole in the center of the ruler. Pull the string through until the portion hanging down below the ruler is about 8–9 cm (3–3¾ in.) long. Secure the string to the top of the ruler with the masking tape. Sprinkle the black pepper over the surface of the water in the bowl.

4. Have one person stir the water counterclockwise with the spoon several times. When the water rotates with a steady spin, remove the spoon.

5. While the water is still spinning, have your partner quickly lower the paper clip about 1 cm (½ in.) into the water, and about 2 cm (1 in.) from the center. Observe the motion of the paper clip. Describe what happened.

_____

6. Stir the water again. This time carefully lower the paper clip into the center of the spiraling water. You may need to try this several times in order to get it right in the center. Make sure to stir the water each time. What happened to the paper clip when it was exactly in the center of the swirling water?

_____

_____

7. In which area did the paper clip exhibit less motion? _____

8. What area of the hurricane did Step 5 represent? _____ Step 6? _____

9. Observe the illustration of a hurricane that your teacher has on display. Compare and describe the motion of the air in the eye wall with the eye of a hurricane.

_____

_____

_____

_____

Name _____

Finding the Epicenter of an Earthquake

Seismic waves recorded on seismographs are used to locate the epicenter of an earthquake. Since P waves arrive first, with S waves following close behind, the difference between the two arrival times are used to determine the distance from the seismograph station to the epicenter. A circle is then drawn on a map, using the distance as the radius and the station as the center of the circle. One circle must be drawn from each of three different stations. The point at which all three circles intersect is the location of the earthquake's epicenter. Follow the directions below to find an epicenter's location.

1. The data table below shows the difference between P and S wave arrival times for three seismograph stations. Use the graph on the next page to locate the distance from the epicenter to each of the three stations. Find the difference in arrival time for Denver on the vertical, or *y*-axis, and follow this line across to the point of intersection with the curved line. Find the distance to the epicenter by reading down from the point of intersection to the horizontal, or *x*-axis. Enter this distance in the *Distance to epicenter (km)* column on the data table below.

2. Repeat *Step 1* for Salt Lake City and Seattle. Record the distances in the table below.

3. Set your compass at a radius equal to the distance from Denver to the epicenter. To do this, use the map scale located on the map of the United States on the next page. Place the point of the compass on the origin of the scale at *0*. Open the compass and set the pencil tip on the distance in kilometers. This is the radius of the circle.

4. Set the point of the compass on Denver and draw a circle around that city using the radius determined in *Step 3*. Make sure to draw the circle carefully.

5. Repeat *Steps 3* and *4* for Salt Lake City and Seattle.

Seismograph station	Difference in arrival times of P and S waves	Distance to epicenter (km)
Denver, Colorado	2 min, 10 sec	
Salt Lake City, Utah	1 min, 10 sec	
Seattle, Washington	1 min, 40 sec	

Name _____

Finding the Epicenter of an Earthquake, continued

Seismic Wave Arrival Times

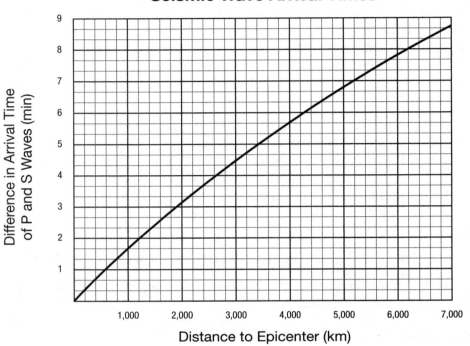

Distance to Epicenter (km)

Name _____

Analyzing the Epicenter

Use the data from the previous two pages to answer the following questions:

1. Observe the three circles you have drawn and locate the epicenter of the earthquake by finding the point of intersection of all three circles. Mark that point with a star. Which one of these three cities is closest to the earthquake's epicenter? _____

_____

2. How far in kilometers is this city from the epicenter? _____

3. In which of the three cities would the earthquake be detected first by the seismographs?

_____

4. Which city would detect it last? _____

5. Where is the epicenter located? _____

_____

6. Approximately how far in kilometers is the epicenter from San Francisco?

_____

7. What would be the difference in arrival times of the P and S waves for a seismograph station in San Francisco? Use the graph to help you with this.

_____

8. When trying to locate an epicenter, why is it necessary to know the distance from the epicenter for at least three seismograph stations? _____

_____

_____

_____

_____

9. Look at the graph. What happens to the arrival times between the P and S waves as the distance from the epicenter increases? _____

_____

_____

Name _____

Building Damage

Read the damage descriptions below. Rate the earthquake's magnitude and level of damage using both the Mercalli and Richter scales for each scenario. Use **BLM 11.2A Mercalli and Richter Scales** to help you find the ratings. Record them on the chart. Answer the questions.

Description of Damage	Mercalli Scale	Richter Scale
Very few people even felt the ground shaking; no damage was reported.		
Bridges were destroyed as a result of this earthquake. The ground split open in many places and very few man-made structures remained standing.		
A few objects were knocked off of shelves and broken. Some people were able to just barely feel this earthquake.		
Houses were moved off of their foundations and other structural damage took place. Trees were uprooted and damaged, and the ground was cracked in some places.		

1. How are the Mercalli and Richter scales alike? _____

_____

2. How are they different? _____

_____

_____

3. Which scale is more scientifically measurable? _____

4. Which scale depends on the observations of people and the damage done to objects?

_____

5. Which scale do you prefer? Why? _____

_____

Name _____

Fire Story

Read and complete the exercises below.

1. Label the three sides of the fire triangle below.

2. Read the story below. While you are reading, keep in mind the three main requirements for a fire that you just labeled above. After reading the story, go back through it and make the following marks regarding either igniting or extinguishing the fire:
- Circle any reference to oxygen.
- Underline any fuel sources that were used.
- Put a box around anything that refers to heat sources.

Reggie and his father went camping in a forest for the weekend. When they arrived at the campsite, they set up their tent and gathered some wood for a campfire. They used small twigs, some crumpled paper, and matches to get their fire burning. For dinner, they cooked over the campfire. As they got ready for bed, they noticed that the campfire was still glowing. Reggie blew into the fire pit and the coals grew hotter and brighter. He put a small twig inside the fire pit and it immediately caught fire. Reggie and his father knew that they should put out the fire before going to bed. First, they tried covering up the fire by placing a metal lid over the fire pit. When they lifted it up a few minutes later, the fire started right back up. Next, they tried spreading out the ashes, but they were tired and did not want to wait for them to burn out. So, they walked to a nearby creek, filled a pail with water, and doused the fire until it was out. Finally, they could get some rest! It had been a long day.

3. Think about what would have happened if Reggie and his father did not put out the fire and it spread to the forest. Write at least four complete sentences explaining how wildfires can affect life in a forested area.

_____

_____

_____

_____

_____

_____

Name _____

Wildfire Damage

The table below shows the number of acres burned by wildfires in the United States between the years 2000 and 2006. Using the data given, draw a pictograph that represents the information in the table. The year 2000 has been done as an example. Once you have completed your pictograph, answer the questions.

Year	Acres burned by wildfires
2000	7.4 million
2001	3.6 million
2002	7.2 million
2003	4.0 million
2004	8.1 million*
2005	8.7 million
2006	9.9 million

*2004 data does not include fires in North Carolina

KEY: ⬤ = 1 million acres of land burned

2000	🔥 🔥 🔥 🔥 🔥 🔥 🔥 (
2001	
2002	
2003	
2004	
2005	
2006	

1. Which year experienced the least number of acres burned by wildfires? _____

2. Which year experienced the greatest number of acres burned by wildfires? _____

3. What is the average number of acres burned per year for this seven-year span? _____

Name _____

Vocabulary Review

Use the words in the Word Bank to fill in the blanks in the paragraphs below. Write the correct answers on the lines. Each vocabulary term is only to be used once.

Word Bank				
storm surge	epicenter	Fujita scale	cinder cone	focus
P waves	tropical cyclone	hurricanes	tornadoes	composite volcanoes
S waves	squall line	vent	shield volcanoes	

The earth is constantly changing as many different systems interact with each other. When the earth's lithospheric plates collide, earthquakes may take place. Earthquakes generate seismic waves, which travel through all layers of the earth. The _____ travel the fastest. Next, the _____ follow close behind, shaking the ground back and forth. People on the earth's surface feel earthquakes at the _____, which is directly above the location where they occur, which is called the _____.

Volcanoes also erupt when systems are out of balance. The magma that pools under the surface in a magma chamber travels through the pipe and then leaves through a _____. Volcanoes are classified into three basic types. A _____ is made from pyroclastic material. Quiet eruptions of thin lava form _____, but _____ are composed of alternating layers of pyroclastic material and lava. The weather also plays an important part of keeping the earth's systems balanced.

A thunderstorm results from changes in warm and cold fronts, which cause instability in the atmosphere. Sometimes a _____, or long line of thunderstorms, can form. When this happens on land, meteorologists look for _____, which can be measured using the _____. When thunderstorms occur over water, they can merge and begin rotating around each other, creating a _____. In the Atlantic Ocean these storms are known as _____. When the storm comes ashore, a wall of water called a _____ follows, which usually does the most damage. A person in the eye of a hurricane might be deceived into thinking that the storm is over because it is so calm there. However, it is only half over!

Name _____

Chapter 11 Review

1. Match the correct term in the box with each description below. Write the corresponding letter on the blank. Each letter will be used more than once. Some terms may have more than one possible answer, but give the answer that fits most accurately.

> T = Thunderstorm and tornado
> H = Hurricane
> E = Earthquake
> V = Volcano
> W = Wildfire

_____ usually associated with squall lines and lightning

_____ usually starts as several thunderstorms, but turns into a much larger storm

_____ requires both warmth and moisture in order to grow

_____ requires warm, humid air to rise rapidly in order to form

_____ usually results from a transform plate boundary that moves

_____ has three types of activity levels—active, dormant, and extinct

_____ is a natural part of the forest's cycle and allows for succession to occur

_____ has three main parts—eye, eye wall, and rain bands

_____ epicenter can be identified using the data of three seismograph stations

_____ holds magma in a chamber underground until pressure triggers the magma to move through the pipe and out the vent

_____ has three types based on formation—shield, cinder, and composite

_____ main damage is done by the storm surge

_____ main damage is caused by extremely high-speed, rotating winds

_____ occurs when oxygen, heat, and fuel come together

_____ gives off body and surface waves, which can be measured

Name _____

Chapter 11 Review, continued

Complete the following diagrams. Write the correct terms on the lines provided.

2. On the three sides of the fire triangle, write the requirements for a fire to occur.

a. _____

b. _____

c. _____

3. Label the following items on the diagram of a volcano: *magma chamber, lava, vent, pipe, crater, ash and dust.*

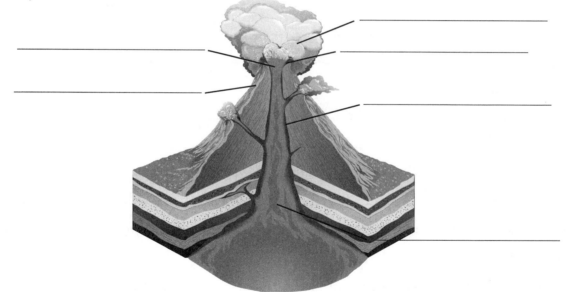

4. On this cross section of the earth, label the *epicenter, focus, body waves,* and *surface waves.*

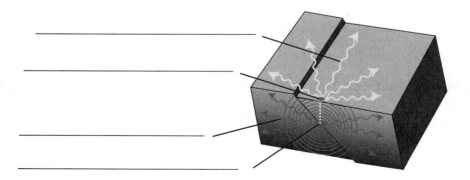

Name _____

Chapter 11 Review, continued

Answer the following questions.

5. How are volcanoes rated? _____

6. What is the name of the scale used to rate hurricanes? _____

How many categories of hurricanes are there? _____

7. What four conditions can sometimes indicate a tornado forming?

a. _____ **b.** _____

c. _____ **d.** _____

8. Describe how to find the epicenter of an earthquake. Be specific.

_____

_____

_____

_____

_____

_____

9. Give one example of something that might trigger a volcanic eruption.

_____

10. Of the two scales that are used to measure earthquakes, which measures the magnitude on a

seismograph? _____

11. List one benefit that wildfires provide for the organisms living in forests.

_____

12. What single part of a hurricane is the most dangerous? _____

What effect causes a great deal of destruction on land? _____

13. Which type of seismic wave is the most destructive? _____

14. What are the three categories of volcanoes based on eruption activity?

a. _____ **b.** _____ **c.** _____

Name _____

Bird's-eye View

Perspective affects the way we interpret the world around us. From physical objects to our hearts and minds, perspective influences our attitudes and beliefs. The earth's position in the universe offers a unique view of the heavens. Follow the directions and complete the exercises below to illustrate the relationship between position and perspective.

1. Your teacher has set up a model of the solar system. The large jug in the middle represents the sun and the bottles represent the planets in orbit. Approximately 2 m (2 yd) from the outermost bottle, lie facedown on the ground and observe the solar system model. If you cannot lie down, move so that your eye level is as close to the ground or floor as possible. In the box below, draw what you see from this position.

2. Staying in the same spot, kneel down. Carefully observe the model. Draw what you see. Make sure to keep the correct angle and perspective of the bottles in your drawing.

3. Staying in the same place, stand to your feet. Note any changes due to the perspective you have now. Draw what you see.

Bird's-eye View, continued

4. Step back from the model so that you are about 10 m (10 yd) away. Remain facing the same direction as before, changing your distance only. Observe the model solar system. Notice what happens to the size of the planets and sun. Draw what you see, keeping this size difference in mind.

5. Move to the raised platform that your teacher directs you to. Observe the model now. Think about how this view affects what you see and show this in your drawing below.

6. Imagine you are a bird flying about 1 km (0.6 mi) overhead. Draw what you think the solar system model looks like from this perspective.

Name _____

Identifying Stars

The following chart is a list of commonly studied stars. Their approximate surface temperatures are rounded to the nearest thousand and the absolute magnitudes are as shown. Use this data to place the *Label* letter for each star in the appropriate location on the H-R diagram on the next Science Notebook page. The first two are already plotted as examples.

Label	Star Name	Type of Star	Approx. Surface Temp. (K)	Absolute Magnitude
a	Sirius A		9,000	+1.4
b	Arcturus		4,000	+0.2
c	Vega		10,000	+0.6
d	Capella		6,000	+0.4
e	Rigel		15,000	−8.1
f	Procyon A		7,000	+2.6
g	Betelgeuse		3,000	−7.2
h	Altair		8,000	+2.3
i	Aldebaran		4,000	−0.3
j	Spica		22,000	−3.2
k	Pollux		5,000	+0.7
l	Deneb		9,000	−7.2
m	Sirius B		25,000	+11.33
n	Achernar		14,000	−1.3
o	Antares		4,000	−5.2
p	Procyon B		8,000	+13.0
q	40 Eri B		17,000	+12.0

Name _____

Identifying Stars, continued

After plotting each star, identify the four populations of stars on the H-R diagram by drawing a circle around each population. Then, determine the type of each star by comparing its location on the H-R diagram to the information below. Write the correct identification for each star—*White Dwarf, Main Sequence, Giant,* or *Supergiant*—in the *Type of Star* column on the chart of the previous Science Notebook page.

White Dwarf: very small stars with fairly high temperature and low absolute magnitude
Main Sequence: small- to medium-sized stars; contains the majority of stars, arranged
along a diagonal line in the middle of the H-R diagram
Giant: large size with fairly low temperature and high absolute magnitude
Supergiant: largest stars with low-to-medium temperature and high absolute magnitude

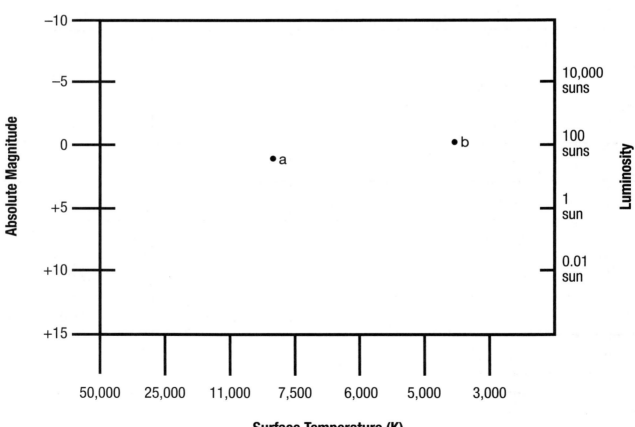

H-R Diagram

Name _____

Changes in the Stars

Do stars transform into other stars just as rocks transform into other rocks? Astronomers' observations and the sciences of chemistry and physics suggest that the cycling of stars could take place. Yet the duration of time necessary for star cycles is much longer than that required for rock cycles. Fill in the blanks to complete the life cycle of a star. Remember that this is a theory of star beginnings and transformations, not a law or principle.

A star begins as a part of a _____, which is a cloud of

_____ and _____. In the most dense part of the

cloud, gravity pulls the matter together and a _____ forms. When this

becomes so dense and hot that nuclear _____ starts, a star is formed.

During the stage known as the _____ _____, a star

burns its fuel and produces _____. The length of this stage depends

on the mass of the star. Interestingly, the more the mass, the faster the star burns and the

shorter its life span. Therefore, a star with low _____uses its fuel more

slowly and lives longer.

A small- to medium-mass star expands into a _____ when it

begins to run out of fuel. As it continues to expand, the outer layers drift away to form

a _____ . The hot core continues to glow from leftover energy and

becomes a _____ _____. Once this star stops

glowing, it is known as a _____ _____. A very

high-mass star becomes a _____. This huge star may continue to

expand to form a _____ , or it may explode, causing a brilliant flash in

an event known as a _____. When the remaining particles in the

dense core lose their charge and become neutrons, it is called a

_____ _____. An extremely massive star may

collapse into a _____ _____ because the gravity

is so intense that even _____ cannot escape. Some of the material

from the supernova expands into space and becomes a _____,

beginning the cycle all over.

Name _____

Star Story

Label the possible stages of a star's life cycle in the diagram below. Use all the terms from the Word Bank to label each stage. Terms may be used more than once. Include brief descriptions of the different stages.

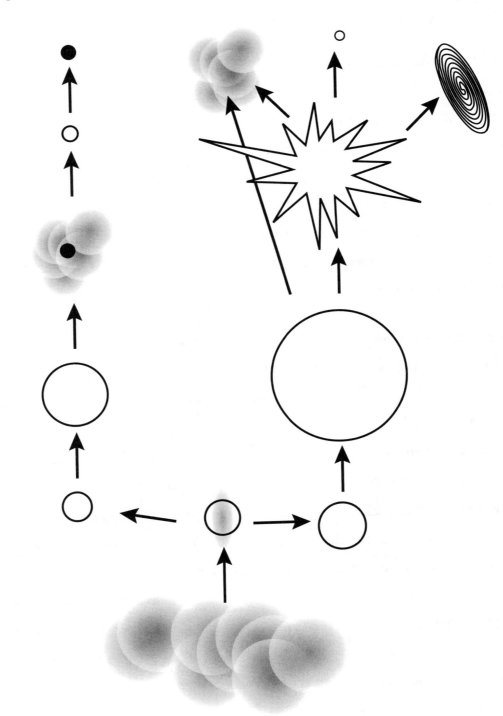

Word Bank
main sequence	white dwarf	giant	supergiant	nebula	supernova
black hole	neutron star	black dwarf	protostar	fusion	

Name _____

Analyzing Theories

Read the following paragraphs about several famous astronomers. Answer the questions below.

In approximately 140 A.D. the Greek astronomer Ptolemy developed a theory known as the *Ptolemaic System*. His theory was based on ideas from Hipparchus (190 B.C.–120 B.C.), who believed in a geocentric, or earth-centered, model of the solar system, and Aristotle (384 B.C.–322 B.C.), who suggested that the planets had circular orbits. Ptolemy also explained the apparent backward motion of the planets by adding that they, as well as the sun, moved in small circles around larger circles known as *epicycles*. This theory was widely accepted for about 1,500 years.

Polish astronomer Nicolaus Copernicus renewed a theory in 1543 that Aristarchus, a Greek astronomer, had first introduced sometime before 230 B.C. They believed in a heliocentric, or sun-centered, model. Copernicus advanced the idea that the orbits of the planets were concentric circles—a nest of circles within larger circles. Ptolemy's idea of epicycles was also adopted by Copernicus. People were hesitant to accept the heliocentric system until Galileo (1564 A.D.–1642 A.D.) collected enough evidence that convinced them it was valid.

About half a century later, Johannes Kepler analyzed some observations made by Tycho Brahe (1546 A.D.–1602 A.D.). At first he assumed circular orbits, but he soon found that his calculations did not fit the observations. Kepler eventually discovered that the orbits were elliptical in shape.

1. Was any part of Ptolemy's theory correct? Explain. _____

_____

_____

2. What idea did Copernicus have that was later found to be true? _____

What beliefs of his were not true? _____

_____

3. Of the astronomers mentioned, who first suggested the geocentric model?

_____

4. Of the astronomers mentioned, who suggested that the sun was the center of the solar system

instead of the earth? _____

5. Who proved that the heliocentric model was correct?

_____

6. What did Kepler determine?

_____

Name _____

Prove It

As observed from Earth, stars appear to move across the sky. Plan and perform an experiment that shows why this apparent motion takes place. You will be given a large flashlight and a ball. You may add any materials to your experiment that you wish, but they must be available to use when testing your plan.

Question: Why do the stars appear to move across the sky?

Predict: _____

_____

Try It Out:

Materials: _____

Procedure: _____

_____

_____

_____

_____

_____

_____

Analyze and Conclude:

1. Did your plan prove your prediction correct? _____

2. If yes, explain what you did during the experiment that showed why the stars appear to move.

_____

_____

_____

3. If not, find a group whose plan worked and try it out. Explain why your plan did not prove why the stars have apparent motion and why the other group's plan did.

_____

_____

_____

Name _____

Telescope Technology

Research the following telescopes and fill in the table to complete the descriptions of each one.

Telescope name or type	Year Completed	Size of collecting lens, dish, and/or array	Type of electromagnetic energy detected (check)				One or two interesting facts about this telescope
			Radio	Infrared	Optical	X-ray	
refracting telescope	1608	various					
reflecting telescope	1668	various					
Lord Rosse's reflector	1845						
radio telescope	1931	various					
Hale Telescope Palomar Observatory	1948						
Infrared Astronomical Satellite (IRAS)	1983						
Hubble Space Telescope (HST)	1990						
Keck I & Keck II Telescopes Keck Observatory	1993, 1996	___ diameter ___ segments					
Chandra X-ray Observatory	1999	_____ mirrors _____ length					
Spitzer Space Telescope (SST)	2003						

Name _____

Predicting Stages

Use the H-R diagram and completed **Science Notebook 12.3A Identifying Stars** to determine the present stages of the stars listed below. Remember that the type of star is also the stage it is in. Then using the information you have learned about the life cycles of stars, predict the next stage of each star.

Star	Present Stage	Next Predicted Stage
Sirius A		
Rigel		
Sirius B		
Betelgeuse		
Capella		
Vega		
Procyon B		
Pollux		

1. Explain how you predicted what the next stage of each star was going to be.

_____

_____

2. What data would have been helpful in specifically determining the next stage of the main sequence stars? Explain why.

_____

_____

3. What other stage or phenomenon can you predict using the same data you suggested in Question 2? Explain how that information helps in your prediction.

_____

_____

Name _____

My Nighttime Sky

The view of the nighttime sky is different depending on your location on Earth. A planisphere is a tool that can help you determine which constellations you should be able to see in your area. Follow the directions below to make your own planisphere. Then use the planisphere to observe and identify the constellations in your nighttime sky.

1. Obtain a manila folder, scissors, and one copy of **BLM 12.7B Star Finder.** You will also need one copy of **BLM 12.7C Northern Sky Wheel** and **BLM 12.7D Celestial Sphere**.

2. Place the star finder from BLM 12.7B onto the half piece of folder. Line up the left side of the star finder with the folded edge of the folder. Glue the star finder onto the folder, making sure to keep the left edge aligned with the fold.

3. Cut along all the edges of the star finder except for the folded edge. Then cut out the inner white oval, being careful to only cut the top side of the folder, not the bottom side.

4. Cut out the sky wheel from BLM 12.7C.

5. Place the sky wheel inside the star finder to complete the planisphere. Turn the wheel until the middle of the current month lines up with the arrow on the time you plan to observe the stars.

6. Hold the planisphere facing down over your head so that you can read it. Point *North* on the planisphere toward the north. Use a compass to determine where north is, if necessary. Look at the stars that appear in the opening of your planisphere. A constellation that appears in the middle of the planisphere should be directly above you. The stars toward the front of the planisphere are in front of you and the stars toward the back of the planisphere are behind you.

7. Look at the night sky and find as many constellations as you can. Match them with the ones on the planisphere.

8. On the celestial sphere from BLM 12.7D, use a fine point, light-colored marker to highlight each constellation that you can find. Make sure to only highlight those that you find.

9. Label each constellation that you have highlighted with the correct name. Use the star wheel to help determine the names of the constellations.

Name _____

Sky Observation Log

After making your planisphere and observing the night sky, complete the exercises below by filling in the correct information.

Observation date: _____   Observation time: _____ to _____

Hemisphere: _____   Location: _____

Latitude: _____   Longitude: _____

Temperature at observation time: _____   Cloud cover (estimate %): _____

1. Describe in general what you saw when looking in the sky. Keep these questions in mind when you write your answer. *How clear was the sky? Did city lights interfere with the visibility? Was the sky full of stars or were only a few visible?* _____

   _____

   _____

   _____

2. List the names of the constellations that you were able to see. _____

   _____

   _____

3. Which constellations would you like to find on another night that you were not able to find

   this time? _____

   _____

   _____

4. Did you observe any other objects in the sky such as airplanes, satellites, comets, or planets?

   If so, list them. _____

   _____

5. Choose one of the constellations that you spotted. Research it and record five facts about it.

   a. _____

   b. _____

   c. _____

   d. _____

   e. _____

Name _____

Vocabulary Review

Write the number within each star above the telescope with the matching description.

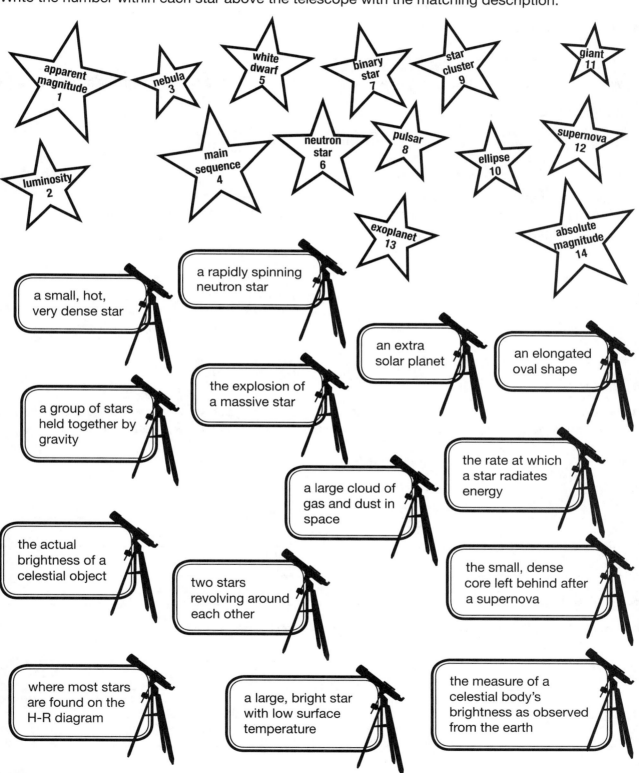

apparent magnitude
1

nebula
3

white dwarf
5

binary star
7

star cluster
9

giant
11

luminosity
2

main sequence
4

neutron star
6

pulsar
8

ellipse
10

supernova
12

exoplanet
13

absolute magnitude
14

a small, hot, very dense star

a rapidly spinning neutron star

an extra solar planet

an elongated oval shape

a group of stars held together by gravity

the explosion of a massive star

the rate at which a star radiates energy

the actual brightness of a celestial object

a large cloud of gas and dust in space

the small, dense core left behind after a supernova

two stars revolving around each other

where most stars are found on the H-R diagram

a large, bright star with low surface temperature

the measure of a celestial body's brightness as observed from the earth

Name _____

Chapter 12 Review

Complete the following exercises:

1. List three factors that affect a star's apparent magnitude.

a. _____ b. _____ c. _____

2. Which of the above factors is adjusted to be the same so that absolute magnitude can be

compared? Explain how. _____

_____

_____

3. Draw and label the temperature and absolute magnitude axes of the H-R diagram. Insert increments on each axis. Circle and label the locations of the four major star populations.

4. Describe the life cycle of a star. Use the following terms in your answer: _main sequence_, _giant_, _supergiant_, _nebula_, _white dwarf_, _supernova_, _black dwarf_, _neutron star_, _black hole_.

_____

_____

_____

_____

_____

5. What is proper motion? _____

6. How will proper motion affect the constellations seen in the future?

_____

_____

7. Explain the difference between an open star cluster and a globular star cluster.

_____

_____

Name _____

Organs to Systems

Place each of the following terms in the appropriate body system category.

kidney	esophagus	platelets	capillaries
biceps	bones	bone marrow	large intestine
heart	ventricles	enzymes	saliva
nerves	ligaments	spinal cord	tendons
mouth	lungs	arteries	stomach
cerebrum	skull	red blood cells	larynx
bladder	diaphragm	urine	nose
aorta	trachea	small intestine	cerebellum
ureters	plasma	white blood cells	vertebrae
atria	veins	brain stem	

Human Body Systems					
Circulatory System	**Respiratory System**	**Digestive System**	**Musculo-skeletal System**	**Nervous System**	**Urinary System**

Name _____

Organ Fitting

Cut out the organs on **BLM 13.1B Body Organs**. Arrange them in their proper locations onto the body outline below. Keep in mind that this is a front view, so some organs will be placed underneath other organs. Glue them into place.

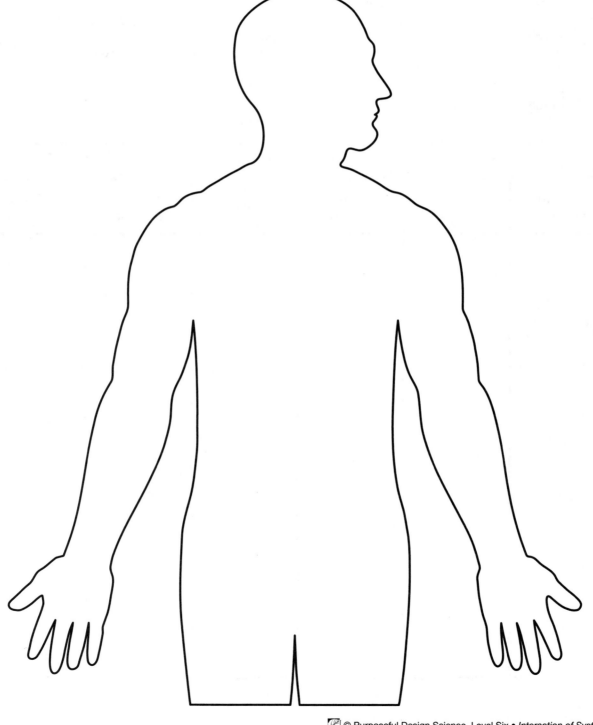

Name _____

Muscle Mania

Read and complete the exercises below.

1. Match the correct type of muscle tissue with the body part. Write *SM* for smooth, *C* for cardiac, and *SK* for skeletal muscle. Use the **Muscular System poster** for help.

_____ stomach   _____ deltoid   _____ heart

_____ rectus abdominis   _____ artery   _____ small intestine

2. Identify each muscle or muscle-lined body part according to its muscular movement type, using *V* for voluntary and *IV* for involuntary.

_____ stomach   _____ heart   _____ smooth

_____ biceps   _____ skeletal   _____ bladder

3. Explain how skeletal muscles work to move the bones.

_____

_____

4. How is the muscular system dependent on the nervous system?

_____

_____

5. _____ keeps food and liquids moving through the digestive tract.

6. Muscles are attached to bones by _____.

7. Even though the diaphragm is a muscle, it is part of the _____ system because it causes the _____ to fill with _____ and to deflate.

8. Look at the picture of the skeletal muscle in your textbook. Explain why blood vessels run throughout the bundles of muscle fibers.

_____

_____

9. Explain the importance of muscular tissue in the digestive system.

_____

_____

Name _____

Muscle Contraction

Question: What happens to skeletal muscle when it contracts continuously over a short period of time?

Predict: _____

Try It Out:
1. Designate who will be the first person to time the activity.
2. For the person who is not the timer, place your arm down by your side. Turn your palm so it faces forward.
3. Grip a textbook that is somewhat heavy, but still manageable.
4. Keeping your upper arm close to your body, raise and lower the book. Only move your lower arm as if you are pumping your arm up and down. Do this steadily for three minutes.
5. Count aloud while your partner times you. Have your partner record the number of times you raise and lower the book each minute for three minutes. Raising and lowering the book equals one time.
6. Switch roles and repeat the activity.

Time	Number of Arm Raises
1 minute	
2 minutes	
3 minutes	

Analyze and Conclude:

1. Name the pair of muscles that are being exercised in this activity.

   a. _____ b. _____

2. Which type of muscle tissue are these? _____

3. Is this type of muscle movement voluntary or involuntary? _____

4. Based upon your observations, what happens to this type of muscle when it contracts repeatedly over a short period of time? _____

5. Would this type of muscle be appropriate for digesting food? Explain.

   _____

   _____

6. Why do you think the heart is not made of this type of muscle?

   _____

Name _____

Bone Decalcification

Read **BLM 13.4B Bone Information**. Then complete the following activity.

Question: What will happen to a chicken bone when placed in vinegar over a five-day period?

Predict: Predict what will happen to the chicken bone in each jar after five days.

Jar #1 (distilled water): _____

Jar #2 (vinegar): _____

Try It Out:

1. Obtain materials from your teacher. Label your jars with #1 and #2 and the initials of the students in your group.
2. Observe the color and flexibility of the two bones. Record your observations on the chart below.
3. Fill Jar #1 half full with distilled water. Fill Jar #2 half full with vinegar.
4. Carefully, place a bone in each jar and put the lids on. Place jars in the designated area.
5. After five days, remove the bones using the forceps. Gently rinse each bone in a cup of water and dry it on a paper towel.
6. Observe and record what happened to the color and flexibility of each bone.

		Day 1	Day 5
Jar #1 (water)	color:		
	flexibility:		
Jar #2 (vinegar)	color:		
	flexibility:		

Name _____

Bone Decalcification, continued

Analyze and Conclude:

1. Bone is made of which type of tissue? _____

2. Why is bone harder and more rigid than cartilage?

_____

_____

3. Why is eating dairy products good for bone health?

_____

_____

4. What substance do specialized bone cells secrete to break down bone tissue?

_____

5. Which substance played this role in your investigation?

_____

6. What was the purpose of Jar #1?

_____

7. Explain the results of your chicken bones after five days.

Jar #1: _____

_____

Jar #2: _____

_____

8. What happens to living bones that are decalcified?

_____

9. Why do you think some areas of your body, such as your nose and ears, remain cartilage
   instead of becoming bone?

_____

Name _____

Bundle of Nerves

Read and complete the following exercises.

1. Listed below are functions and actions that require a response from the brain. Determine whether the cerebrum (C), cerebellum (L), or the medulla oblongata (MO) is responsible for performing that action. Write the letter for the correct answer on the blanks below.

_____ Mom asks you to run to the store and buy some milk.

_____ You remembered to do it later that day.

_____ As you were running, your heart rate increased.

_____ You spotted the milk in the cooler at the back of the store.

_____ After drinking the milk on the way home, your breathing increases.

_____ Tired from running, you slowed to a walk.

_____ While walking home, you read the back of the milk carton.

2. Draw a neuron and label its parts.

3. What makes up the central nervous system (CNS)? _____

4. What makes up the peripheral nervous system (PNS)? _____

_____

5. Describe a nerve. _____

_____

6. Explain how the nerves of the PNS process messages without involving the brain. Give

an example. _____

_____

_____

7. Do you think a reflex action is a voluntary or an involuntary response? Explain why.

_____

_____

Name _____

Brain Protection

Read the following information:

Mankind is set apart from every other organism on Earth due, in part, to the human brain. This part of the central nervous system has the capacity to think in a logical way, make decisions, and direct the process of writing and talking. Therefore, it is no surprise that the soft, wrinkled brain with a gelatin-like consistency must be supported and protected. The hard bony cranium that covers the brain is the first line of defense. Additionally, the brain is surrounded by three layers of tough membranes called *meninges*. Fluid fills in any spaces between the meninges and open spaces within the brain itself. This cushions the brain from being injured when an object or surface comes in contact with the skull. This same fluid is found in the vertebral column, buffering the spinal cord from injury.

Follow the directions below to simulate the protection of the brain and spinal cord by the skull and vertebral column, meninges, and the cerebrospinal fluid:

1. Gently place a raw egg in its shell in an empty jar. Cap the jar and shake it. What happens?

_____

What would happen if the brain were unprotected inside the skull in a similar way?

_____

2. Fill a second jar with water and gently place another raw egg in its shell into the jar. Cap the jar and shake it. What happens this time? _____

_____

What part of the of the brain's protective devices does this represent?

_____

3. Remove the unbroken egg from the jar. Gently place the egg in a small sealable bag of water and seal the bag. Place the bag in the jar of water. Cap the jar and shake it. How is this different from *Step 2*? _____

_____

What part of the brain's protection system does this resemble? _____

_____

4. Why do you think it is important to wear a helmet while biking, skiing, or playing certain sports?

_____

_____

Name _____

Owl Pellet Dissection

Owls are birds of prey that kill and eat other animals. Typically they eat small mammals such as rats, mice, shrews, and moles. Owls have a hooked beak and can tear their prey in pieces or swallow it whole. However, the bones, feathers, and fur are not digested. Instead, they are regurgitated by the owl in the form of a pellet. The diet of a particular owl can be discovered by collecting its pellets and dissecting them. Follow the directions below to dissect an owl pellet.

1. Put on a mask and plastic gloves before handling the owl pellet. Inspect the outside of the owl

pellet and measure its length and width. Record your observations. _____

_____

_____

2. Place the pellet in the bowl. Using the eyedropper, add enough water to wet the pellet. Use the forceps and dissecting needles to carefully pull the pellet apart. Separate the bones from the fur and any other body parts.

3. Clean off the bones by gently swirling them in some water. Place the bones onto a paper towel and carefully dry them off.

4. Use the rodent bone X-ray below as a guide to sort the bones. Reconstruct the skeleton on a piece of construction paper and glue the bones in place. Keep in mind that owls usually eat more than one rodent, so you may have multiple skeletons.

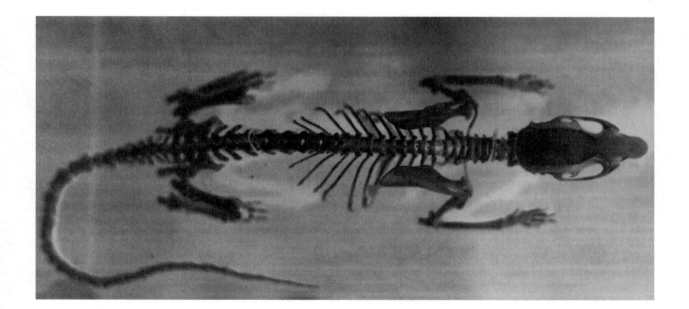

Name _____

Surface Area

Villi greatly increase the surface area of the small intestine. Perform the experiment below, using the materials your teacher provides, to determine the effect of increased surface area.

Question: How does the amount of surface area affect the rate of absorption in the digestive tract?

Predict: _____

_____

Try It Out:

1. Place one sugar cube into a sealable plastic bag. Gently crush it until it is in coarse pieces.
2. Place one sugar cube into a second sealable plastic bag. Crush it until it is in fine pieces.
3. Compare the amount of surface area of the three samples of sugar. Answer Question 1 below.
4. Label cups *1*, *2*, and *3*, and fill each half full with water.
5. Place the whole sugar cube in *Cup 1*.
6. Stir the sugar until it is dissolved. Time how long it takes to dissolve and then record the information on the chart below.
7. Repeat Step 6, using *Cup 2* and the coarse particles of sugar.
8. Repeat Step 6, using *Cup 3* and the fine particles of sugar.

Sugar particles	Time to dissolve (in seconds)
Cup 1 (whole cube)	
Cup 2 (coarse particles)	
Cup 3 (fine particles)	

Analyze and Conclude:

1. What did you determine about the amount of surface area of the three samples of sugar?

_____

2. Which sugar sample dissolved the fastest? The slowest? _____

3. Which sugar sample would absorb the fastest during digestion? _____

Explain. _____

_____

4. How does this relate to the purpose of the villi in the small intestine? _____

_____

_____

Name _____

Sensing

1. In your own words, explain how light passes through the different parts of the eye and converts to impulses that are sent to the brain. Then, label the parts of the eye below.

_____

_____

_____

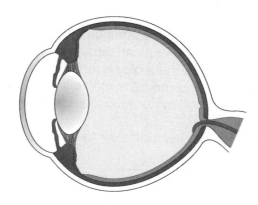

2. In your own words, describe how sound enters the ear and travels to the brain. Label the parts of the ear and draw arrows on the illustration to show the direction that the vibrations take as they move through the ear.

_____

_____

_____

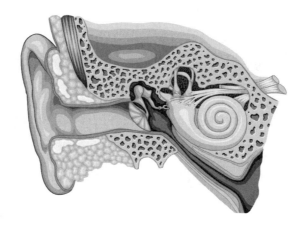

Name _____

Sensory Neurons

Read the paragraphs below and answer the questions.

There are three types of neurons—sensory neurons, interneurons, and motor neurons. Sensory neurons pick up stimuli from both the internal and external environment. They convert this information into a nerve impulse. At the tip of sensory neurons are specially designed receptors. Chemoreceptors detect molecules and are used for taste and smell. Thermoreceptors sense hot and cold temperatures. Mechanoreceptors, used for touch, detect changes in pressure and position. Pain receptors pick up pain sensations due to severe heat and pressure. Photoreceptors absorb light energy.

Nerve impulses travel through the sensory neurons until they reach interneurons. These are found in the spinal cord and the brain. Interneurons can carry impulses from one neuron to another. Once the brain interprets the message, the motor neurons may send impulses to muscles or glands and trigger them to respond. Interneurons also can respond to sensory messages as well as be involved in thought, memory, and motor messages.

1. List the three types of neurons and describe their functions.

a. _____

_____

b. _____

_____

c. _____

_____

2. Write the name of the receptor responsible for detecting each type of stimulus listed below.

_____ **a.** a campfire

_____ **b.** stubbing your big toe

_____ **c.** a cake baking in the oven

_____ **d.** weather with below freezing temperatures

_____ **e.** eating candy

_____ **f.** feeling someone brush up against you

_____ **g.** losing balance while walking on a beam

_____ **h.** watching a thunderstorm

Name _____

Vocabulary Review

Read the following statements and fill in the blanks with the correct vocabulary terms:

1. In regard to surface area, _____ are to the lungs as

_____ are to the small intestine.

2. A _____ joins a muscle to a bone whereas

_____ can function as a cushion between bones.

3. Food moves through the digestive system by the action of _____.

4. _____ is the partially digested food in the stomach and small intestine.

5. Exoskeleton is to invertebrate as _____ is to vertebrate.

Match the definition with the correct vocabulary word.

_____ **6.** pyloric sphincter   **a.** the breastbone

_____ **7.** sternum   **b.** sends impulses from the cochlea to the brain

_____ **8.** neuron   **c.** small, lower portion of the brain

_____ **9.** optic nerve   **d.** sends impulses from the retina to the brain

_____ **10.** medulla oblongata   **e.** found between the stomach and the
small intestine

_____ **11.** auditory nerve   **f.** the basic unit of the nervous system

12. There are 13 vocabulary words in this chapter. List each one under the system or group of organs to which it belongs. Some of the terms can be used in more than one column.

Nervous   Musculoskeletal   Digestive   Respiratory   Sense Organs

Name _____

Chapter 13 Review

Answer the following questions:

1. What happens to the diaphragm when a person exhales?

_____

_____

2. What primarily happens in the following structures:

  a. stomach: _____

  b. small intestine: _____

  c. large intestine: _____

  d. alveoli: _____

  e. capillaries: _____

3. What does the vertebral column protect? _____

4. Cross out the term that least fits with the others. Using the other terms, explain why it does not fit.

  a. cartilage   calcium   biceps   ossification   bone

_____

_____

  b. peristalsis   smooth muscle   food   absorption   digestive tract

_____

_____

  c. brain   nerves   cranium   spinal cord   CNS

_____

_____

5. Explain how four of the body systems are interdependent.

_____

_____

_____

_____

Name _____

Organic Molecules

There are six basic groups of nutrients necessary for maintaining good health—carbohydrates, proteins, fats, vitamins, minerals, and water. Food contains these six kinds of nutrients that help you grow. Some foods contain all six nutrients. When you chew food, you are physically breaking down the portion into more manageable sizes to swallow. This is a type of physical digestion. Once the food is in the stomach, it undergoes more physical digestion. Then, substances in the body called *enzymes* perform chemical digestion.

Your teacher has given you a variety of food samples for you to taste. In the chart below, write the name of the food. Once you have tasted it, record whether you think the main nutrient in it is a carbohydrate, protein, or fat. In the third column, explain how you came to your conclusion. Then answer the questions.

Food Item	Nutrient	Explanation

1. Do the textures or tastes of the food samples give hints to your predictions? Explain.

_____

_____

2. Did you use any particular criteria more than others to determine the types of nutrients? Explain. _____

_____

3. What substances break down nutrients into particles that can be absorbed?

_____

_____

Name _____

Emulsification of Fats

Fats come from animal sources such as butter, lard, and bacon. Other fats, like olive oil and corn oil, come from vegetable sources. Some fats are harder to see, such as those in egg yolks, milk, and nuts. Besides providing more energy within every gram, fats provide food with texture and flavor. They also help reduce hunger pangs between meals due to the body's slow digestive rate of fats. Most digestion of fat begins in the small intestine. Bile is a digestive fluid that is produced in the liver and stored in the gall bladder. When bile enters the small intestine, it breaks larger fat globules into fat droplets through a physical reaction called *emulsification*. A second reaction involves the enzyme lipase. During this chemical reaction, lipase breaks the fat droplets into fatty acids and a substance called *glycerol*, which are then able to diffuse into the cells lining the intestine. Follow the directions below to simulate the emulsification reaction of bile on fat globules in the small intestine. Answer the questions.

1. Fill a test tube half full of water. Add 10 drops of vegetable oil to the test tube.
2. Place a stopper in the tube and gently shake it. Record your observations.
3. Now add 10 drops of dishwashing detergent to the test tube. Replace the stopper and gently shake the test tube. Record your observations.

Test Tube Ingredients	Observations
Water, Oil	
Water, Oil, Detergent	

4. Where in the body does most fat digestion occur? _____

5. What role does the detergent simulate in this activity? _____

6. What type of reaction is this when it occurs in the digestive system? Describe what happens.

_____

7. Why are the liver and the gall bladder important organs for fat digestion?

_____

8. What substance is necessary for fat to be further digested? _____

9. What type of reaction occurs at this stage? _____

10. What are the by-products of fat digestion? _____

11. How many reactions does it take to break fat down into particles that are small enough to be

absorbed into the cells? _____

Name _____

Sheep Heart Dissection

The sheep heart is very similar to the human heart in both size and structure. Study the illustrations of the ventral and dorsal views on **BLM 14.3E Sheep Heart External Anatomy**. Use *Figures A–H* on **BLM 14.3F–G Dissection Stages** to guide you through the dissection process. Be precise with your incisions. Otherwise, you may not obtain the desired results.

1. Put on gloves and place the sheep heart in the dissection pan. Identify the ventral and dorsal

sides. Describe the differences. _____

_____

2. Locate the following blood vessels and areas of the heart. Check off each one as you find it. Use *Figure A* to ensure the proper position of the vessels. Have your teacher check this before continuing.

_____aorta	_____apex	_____coronary artery
_____superior vena cava	_____brachiocephalic artery	_____pulmonary vein
_____pulmonary artery	_____right/left atria	_____right/left ventricles

3. Position the heart with the ventral side showing as seen in *Figure B*.

4. Place the scalpel about 2 cm to the left of the coronary artery and at the base of the pulmonary artery. Measure with a ruler as in *Figure C*. Carefully make a shallow incision running parallel to the coronary artery down as far as you can go without turning the heart over. Remember to stay parallel through the entire incision. Gently open the heart as seen in *Figure D*. Look inside. The pulmonary valve and the tricuspid valve should be visible. Describe

what you see in detail. _____

_____

5. Look at *Figure E* for details. Keeping parallel to the coronary artery, cut around to the dorsal side. Notice the coronary artery curves upward and runs vertically. Stop the incision just to the right of the dorsal coronary artery. Find the superior vena cava (SVC). Staying parallel to the coronary artery, cut downward through the SVC until this incision meets the first incision.

6. Very gently spread the sides apart, as shown in *Figure F*. The whole right interior should be visible now. Describe the difference between the walls of the right atrium (RA) and the right

ventricle (RV). _____

Explain why there is a difference. _____

7. Find the tricuspid valve located between the RA and the RV. What is the purpose of this valve?

_____

_____

Name _____

Sheep Heart Dissection, *continued*

8. Place the heart back on the pan with the ventral side up. Find the pulmonary vein (PV), which is located on the top toward the back of the left atrium, above and to the left of the coronary artery. Make an incision through the PV straight down to the apex. Use *Figure G* as a guide.

9. Gently open the left side of the heart, as shown in *Figure H* and observe the walls of the left ventricle (LV). Describe the difference between the walls of the LV and the RV.

_____

Why is there a difference? _____

_____

10. Look for the bicuspid valve between the LA and the LV. Describe what you see.

_____

11. Find the aorta. Keep the left side of the heart open. Gently slide a probe into the aorta until

you see the tip. Into what chamber does it empty? _____

12. Slide the probe down the pulmonary vein. Where does it go? _____

13. Slide the probe down the superior vena cava. To what chamber does it go? _____

14. Compare your observations in Exercises 11–13 with what you have previously learned about the structure of the heart. Do your observations support this information?

_____

15. What was the most interesting thing you found about this dissection?

_____

_____

_____

_____

16. Observe the interior structure of the sheep heart again. Describe how the heart is perfectly suited for its purpose. Reflect on how this deepens your faith in God the Creator.

_____

_____

_____

_____

_____

Name _____

Bone Design

Question: Do bones have particular designs that allow for strength? To test this, you will be making 3-D shapes out of paper and placing weight on them.

Prediction: Circle one answer for each statement about shapes and how they compare to bones.

I predict a cylinder will be (stronger/weaker) than a rectangular column.

I predict additional sheets of paper will (strengthen/weaken/have no effect on) the ability of the cylindrical column and rectangular column to hold weight.

Try It Out:

1. Roll one sheet of paper lengthwise, leaving a 2.5 cm (1 in.) overlap. Tape it into the shape of a cylinder.
2. Fold another sheet of paper lengthwise, leaving 2.5 cm (1 in.) exposed at the edge of one side. Fold the 2.5 cm (1 in.) exposed portion over and crease it, creating an overlap. Open the paper. Keep the overlap folded down. Fold each lengthwise edge to the middle fold-line and crease the folds. Form an open rectangular column with the overlap on the outside. Tape it into place.
3. Stand both structures on the desk. Use books of equal size and weight. At the same time gently place a book on each of the structures and observe which one collapses first. Record the results on the data chart.
4. Repeat *Steps 1–2*, using two sheets of paper for each structure. Make another one-sheet cylinder and one-sheet rectangular column. Compare the one-sheet cylinder and two-sheet cylinder by performing *Step 3*. Add a second book if necessary. Make sure equal amounts of weight are added to both structures. Record the results.
5. Repeat *Step 3* with the one-sheet rectangular column and the two-sheet rectangular column. Record the results.
6. Repeat *Step 3* for the two-sheet cylinder and the two-sheet rectangular column. Add extra books whenever necessary. Record your results.

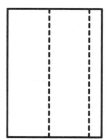
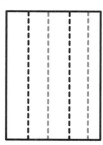
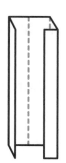

Name _____

Bone Design, continued

Structures	Results
One-Sheet Cylinder / One-Sheet Rectangular Column	
One-Sheet Cylinder / Two-Sheet Cylinder	
One-Sheet Rectangular Column / Two Sheet Rectangular Column	
Two-Sheet Cylinder / Two-Sheet Rectangular Column	

Analyze and Conclude: Answer the following questions:

1. Which structural shape is stronger? _____

2. Does this change when the number of sheets of paper changes?

_____

3. How does the thickness of the structures affect the strength of the structures? Explain for both

the cylinders and both the rectangular columns. _____

_____

_____

_____

4. Based on this experiment, which bone shape would bear more weight?

_____

5. Which general shape do weight-bearing bones, such as leg bones, have?

_____

6. In the experiment, which models represented more dense bones?

_____

7. How does bone density affect the strength of the bone? _____

_____

8. List two ways you can increase bone density and strength at an early age.

a. _____   **b.** _____

Name _____

Impulse Pathway

Complete the following exercises.

1. Identify the *dendrites*, *axon*, *myelin*, and *synapse* on the illustration of a typical neuron below. Draw a line from the specific area and write the term on the line. Then draw arrows to indicate the direction of the impulse pathway from one end to the other.

2. What part of the neuron picks up a stimulus or an impulse? _____

3. Where does the electrical impulse go after traveling through the axon?

_____

4. What specific nutrients are needed for proper transmission of electrochemical impulses?

_____

5. Distinguish between a dendrite and an axon in three different ways.

DIFFERENCES	DENDRITE	AXON
1		
2		
3		

Name _____

Neuron Connections

Read and follow the directions below to perform the activity. Answer the questions that follow.

1. Stand at an arm's length from another student. The entire class should form one straight line. Your teacher will measure the length of the line. Write the length in meters.

_____

2. Your teacher will give you a small object, such as a pebble or eraser. Hold it in your right hand. Everyone should have an object except the last person in line.

3. When your teacher tells you to start, the person at the beginning of the line should place his or her object into the left hand of the next person in line. Once the object has been placed in the second person's hand, that person will in turn place his or her first object in the third person's left hand. Continue this sequential pattern until the last person receives an object. Everyone, except the first person in line, should now have their neighbor's object in their left hand.

4. Return the objects back to the right hand of the person next to you. Start the activity again.

   This time your teacher will time how long it takes. Write down the time. _____

5. How does this activity relate to what you have learned about the nervous system?

_____

6. What part of the neuron in this activity does each of the following represent?

   left hand and arm _____   right arm _____

   person's torso _____   right hand and fingers _____

   space between hands _____   object _____

7. Calculate the rate of transmission of the simulated impulse by dividing the length of the line in meters by the time in seconds that it takes for the object to travel down the line. Show your work and write the answer on the line provided.

   Answer: _____

8. An impulse can travel along a myelinated neuron at the rate of 120 meters/second (m/s). In a non-myelinated neuron, an impulse can travel at about 2 m/s. Compare the transmission rate of your simulated impulse to these two transmission rates.

_____

9. Explain the importance of myelin. _____

_____

Name _____

Nutrient Testing

Obtain materials and listen carefully for directions from your teacher. Complete the exercises.

Question: Which commonly eaten foods and drinks contain lipids (fat), protein, starch, and/or sugar?

Predict: Predict which food samples contain lipids, protein, starch, and/or sugar by placing a check mark in one or more columns.

Foods	Fat	Protein	Starch	Sugar
cracker				
potato				
yogurt (plain)				

Try It Out: Use **BLM 14.6B Directions: Nutrient Testing** to conduct four tests on three food samples. Record whether the food sample is positive (+) for the nutrient or negative (-).

Foods	Fat	Protein	Starch	Sugar
cracker				
potato				
yogurt (plain)				

Name _____

Nutrient Testing, continued

Analyze and Conclude: Answer the following questions:

1. Starch and sugar belong to which group of nutrients?_____

2. What is the basic unit for the following organic molecules?

    a. lipids (fat) _____

    b. protein _____

    c. carbohydrates _____

3. Explain what happens to most foods before they can be utilized by the cells.

_____

_____

4. What happens to nutrients once they reach the cell?

_____

_____

5. Of the foods tested, which ones are sources for at least three of the tested nutrients?

_____

_____

6. Are any of the foods listed in Question 5 good for a healthy diet? Explain. Include any suggestions to help maintain a balanced diet.

_____

_____

_____

7. List other important considerations necessary in establishing a healthy diet besides looking for foods that contain the necessary nutrients.

_____

_____

_____

_____

Name _____

The Meal Connection

Imagine you are a nutrition expert. You have been asked by a school principal to look over three meals that have been suggested to be offered on the menu. Use your knowledge about nutrients and cooking methods to provide the most nutritious, heart-healthy meal possible. Make any changes that you feel are necessary.

Meal 1

fried egg _____

hash browns _____

toast with butter and jelly _____

fruit drink _____

whole milk _____

Meal 2

double patty hamburger _____

bun _____

tomato and lettuce _____

mayonnaise _____

french fries (large order) _____

chocolate chip cookie _____

can of soda _____

Meal 3

grilled cheese sandwich (with butter) _____

fried potatoes _____

fruit roll-up _____

chocolate cake _____

fruity drink made from mix _____

Name _____

Nutrients and Daily Diet

Calories denote how much energy a serving of food will provide. Being aware of the number of calories eaten can help a person manage body weight. In general, Calories for moderately active adolescents should be 1,800–2,000 per day for girls and 2,000–2,200 for boys. These limits are less for those with sedentary lifestyles. Where the calories come from is important for a healthy diet, too. No more than 10% of the daily diet's Calories should come from sugar. Total fat intake should be no more than about 30% of the daily diet. Less than 7–10% of the fat Calories should come from saturated fat. Cholesterol limits should be less than 300 mg per day, no matter how many Calories are consumed.

Calculate the number of Calories per day allowed for sugar, total fat, and saturated fat. Use the general daily Calorie figures for adolescent boys and girls.

Suggested Amounts of Nutrients per Day (in Calories)

Nutrients	1,800 Calories / Day	2,000 Calories / Day	2,200 Calories / Day
Sugar			
Total Fat			
Saturated Fat			

Show your calculations below.

Name _____

Vocabulary Review

A publishing company has asked you to edit the following terms for accuracy. If the definition is correct, write *correct* on the line. If the definition is not accurate, rewrite it to make it accurate.

1. *electrolyte*: a substance that bonds to water and is not a good conductor

_____

2. *carcinogen*: a substance or agent that causes cancer

_____

3. *lipid*: the mineral that includes fats and waxes

_____

4. *calorie*: the sugar content in food

_____

5. *synapse*: a tiny gap between one neuron and another

_____

6. *cholesterol*: a type of protein produced by the kidneys and found in blood

_____

7. *dendrite*: a highly branched microscopic extension of a neuron's cell body

_____

8. *myelin*: a fatty substance that surrounds beef steaks

_____

9. *hemoglobin*: the protein found in white blood cells that forms platelets

_____

10. *hypertension*: the term for high blood pressure

_____

11. *osteoporosis*: a disease caused by obesity

_____

12. *axon*: a single, long projection that extends from the brain

_____

13. *ligament*: the strong, elastic tissue that connects bone to muscle at movable joints

_____

Name _____

Chapter 14 Review

Answer the following:

1. List the six nutrients in a balanced diet.

    a. _____   **b.** _____

    c. _____   **d.** _____

    e. _____   **f.** _____

2. What are three significant benefits of exercise?

    a. _____

    b. _____

    c. _____

3. What is the function of myelin? _____

4. Explain the path an electrical impulse takes as it travels from one neuron to the next.

_____

_____

5. Tell which type of organic compound is represented by the example:

    C = carbohydrate   P = protein   L = lipid

    _____ starch   _____ animal lard   _____ cholesterol   _____ fruit

    _____ egg white   _____ flour   _____ sucrose   _____ fish

6. How are food nutrients used in the body?

_____

7. Why is iron necessary in the diet? _____

8. Distinguish between saturated and unsaturated fats. Give examples of both.

_____

_____

9. Why are sodium and potassium necessary for the body?

_____

10. List three nutrients and explain why the body needs them for a healthy cardiovascular system.

_____

_____